Some notes

ROBERT WENSLEY
Editor: Jason Murgatroyd

Edexcel International GCSE
Chemistry

EDEXCEL CERTIFICATE IN CHEMISTRY

PRACTICE BOOK

Every effort has been made to trace all copyright holders, but if any have been overlooked the Publishers will be pleased to make the necessary arrangements at the first opportunity.

Although every effort has been made to ensure that website addresses are correct at time of going to press, Hodder Education cannot be held responsible for the content of any website mentioned. It is sometimes possible to find a relocated web page by typing in the address of the home page for a website in the URL window of your browser.

Orders: please contact Bookpoint Ltd, 130 Milton Park, Abingdon, Oxon OX14 4SB. Telephone: (44) 01235 827720. Fax: (44) 01235 400454. Lines are open 9.00–17.00, Monday to Saturday, with a 24-hour message answering service. Visit our website at www.hoddereducation.co.uk

© Robert Wensley 2013

First published in 2013 by

Hodder Education

An Hachette UK Company,

338 Euston Road

London NW1 3BH

Impression number	5	4	3	2	1
Year	2017	2016	2015	2014	2013

All rights reserved. Apart from any use permitted under UK copyright law, no part of this publication may be reproduced or transmitted in any form or by any means, electronic or mechanical, including photocopying and recording, or held within any information storage and retrieval system, without permission in writing from the publisher or under licence from the Copyright Licensing Agency Limited. Further details of such licences (for reprographic reproduction) may be obtained from the Copyright Licensing Agency Limited, Saffron House, 6–10 Kirby Street, London EC1N 8TS.

Cover photo © Andrew Brookes/Corbis

Illustrations by Aptara, Inc.

Typeset in ITC Legacy Serif by Aptara, Inc.

Printed in Spain

A catalogue record for this title is available from the British Library

ISBN 978 1 444 179200

Contents

Useful formulae and abbreviations	iv
Get the most from this book	v
Periodic Table	vi

1 Principles of chemistry 1
Practical work	1
Using the Periodic Table	4
Calculations	7

1 Principles of chemistry 2
Charges, chemical formulae and equations	10
Using electronic configurations	12
Data analysis	16

2 Chemistry of the elements
Using the Periodic Table	19
Using electronic configurations	22
Charges, chemical formulae and equations	25
Practical work 1	28
Practical work 2	31
Data analysis	33
Longer-answer questions	36

3 Organic chemistry
Calculations	38
Structure, properties and reactions of alkanes and alkenes	39
Longer-answer questions	43

4 Physical chemistry
Charges, chemical formulae and equations	45
Practical work 1	48
Practical work 2: making salts	52
Data analysis	55
Working with graphs	58
Calculations 1	62
Calculations 2	66

5 Chemistry in industry
Charges, chemical formulae and equations	69
Practical work	72
Data analysis	76
Working with graphs	79
Calculations	83
Longer-answer questions	85

Index
	87

Useful formulae and abbreviations

A_r relative atomic mass

M_r relative formula mass

number of protons = atomic number

number of electrons = number of protons

number of neutrons = mass number − atomic number

$$\text{number of moles} = \frac{\text{mass of element or compound in grams}}{A_r \quad \text{or} \quad M_r}$$

volume of 1 mole of gas at room temperature and pressure (r.t.p.) = 24 dm³ or 24 000 cm³

$$\text{concentration of a solution in mol/dm}^3 = \text{number of moles dissolved} \times \frac{\text{volume of solution in cm}^3}{1000}$$

$$\text{percentage yield} = \frac{\text{mass obtained by the reaction}}{\text{mass expected to be obtained}} \times 100$$

heat energy change (J) = mass of solution (g) × specific heat capacity of solution (4.2 J/g) × temperature change (°C)

mass of 1 cm³ of water = 1 g

1 faraday = 1 mole of electrons

1 mole of electrons = 96 500 coulombs of electric charge

Get the most from this book

This Practice Book will help you to prepare for your International GCSE Chemistry assessment. The questions are arranged in Sections to match the Specification, so that you can use this book throughout the year as you complete each Section, or as part of your final revision.

We have included lots of examples of the types of questions that may be included in your International GCSE Chemistry examinations. The different types of questions cover a large range of topics, giving you an opportunity to check your understanding of the content and requirements for the examinations.

For each question type, you will find a sample question and two different student responses, with comments explaining the correct approach, to help you build good examination techniques.

Remember, in your final examination, questions can consist of a number of different parts, covering the content of more than one Section, so make sure that you revise well and cover all the questions in this book.

All answers are available online at www.hodderplus.co.uk/edexcelgcsescience

> Carefully study **Examples** of exam-style questions and two different sample responses to see how marks are allocated.

> Marker comments and tips for success show you how to prepare and give the best response to help you improve your marks.

> Once you understand how to approach this type of question, have a go at the **Practice questions**.

---

3 Organic chemistry

Calculations

Example

1. A student investigated a compound similar to ethanol. The student found that this compound contained 59.76% by mass of carbon, 13.33% by mass of hydrogen, and that the rest was oxygen.

   Calculate the empirical formula of this compound. (5)

   (Total for question = 5 marks)

Student 1 response — Total 2/5	Marker comments and tips for success
$C, \frac{59.76}{12} = 4.98$ O $H, \frac{13.33}{1} = 13.33$ O So ratio = $\frac{13.28}{4.98}$ = 2.67 or 1:2.67 or 3:8 ✓ So formula is C_3H_8 ✓	Always read the question carefully. This answer does not include oxygen in the calculation. As a result, the suggested formula is correct for carbon and hydrogen, but has no oxygen. On an *error carried forward* the answer gains 2 marks.

Student 2 response — Total 5/5	Marker comments and tips for success
$O = 100 - 59.76 - 13.33 = 26.56\%$ ✓ C is $\frac{59.76}{12} = 4.98$, H is $\frac{13.33}{1} = 13.33$, O is $\frac{26.56}{16} = 1.66$ ✓ Divide by 1.66 ✓ to give $C = 3, H = 8, O = 1$ ✓ Formula is C_3H_8O ✓	1 mark for correctly calculating the percentage of oxygen and 1 mark for dividing by the relative atomic masses. Finding the simple ratio gains 1 mark, with the final mark for the correct empirical formula. You should try to lay out your calculation answers like this one. Allow a line for each step, and clearly state what you are doing.

Practice questions

2. A student investigated a compound similar to ethanol. The student found that this compound contained 37.8% by mass of carbon, 50.4% by mass of oxygen, and that the rest was hydrogen.

   Calculate the empirical formula of this compound. (5)

3. A student investigated a compound containing hydrogen, carbon and oxygen. The compound contained 40.0% by mass of carbon, 6.7% by mass of hydrogen, and 53.3% by mass of oxygen.

   Calculate the empirical formula of this compound. (4)

4. A scientist investigated a hydrocarbon with formula mass 112. The hydrocarbon had 85.7% by mass of carbon and 14.3% by mass of hydrogen.

   a) Calculate the empirical formula of the hydrocarbon. (4)

   b) Calculate the formula of the hydrocarbon. (3)

5. A student investigated a compound. The compound had 56.8% by mass of chlorine, 38.4% by mass of carbon and the rest was hydrogen.

   Calculate the empirical formula of this compound. (5)

Periodic Table

I	II												III	IV	V	VI	VII	0
1	2												3	4	5	6	7	
						1 **H** Hydrogen 1												4 **He** Helium 2
7 **Li** Lithium 3	9 **Be** Beryllium 4												11 **B** Boron 5	12 **C** Carbon 6	14 **N** Nitrogen 7	16 **O** Oxygen 8	19 **F** Fluorine 9	20 **Ne** Neon 10
23 **Na** Sodium 11	24 **Mg** Magnesium 12												27 **Al** Aluminium 13	28 **Si** Silicon 14	31 **P** Phosphorous 15	32 **S** Sulfur 16	35.5 **Cl** Chlorine 17	40 **Ar** Argon 18
39 **K** Potassium 19	40 **Ca** Calcium 20	45 **Sc** Scandium 21	48 **Ti** Titanium 22	51 **V** Vanadium 23	52 **Cr** Chromium 24	55 **Mn** Manganese 25	56 **Fe** Iron 26	59 **Co** Cobalt 27	59 **Ni** Nickel 28	64 **Cu** Copper 29	65 **Zn** Zinc 30	70 **Ga** Gallium 31	73 **Ge** Germanium 32	75 **As** Arsenic 33	79 **Se** Selenium 34	80 **Br** Bromine 35	84 **Kr** Krypton 36
85 **Rb** Rubidium 37	88 **Sr** Strontium 38	89 **Y** Yttrium 39	91 **Zr** Zirconium 40	93 **Nb** Niobium 41	95 **Mo** Molybdenum 42	99 **Tc** Technetium 43	101 **Ru** Ruthenium 44	103 **Rh** Rhodium 45	106 **Pd** Palladium 46	108 **Ag** Silver 47	112 **Cd** Cadmium 48	115 **In** Indium 49	119 **Sn** Tin 50	122 **Sb** Antimony 51	128 **Te** Tellurium 52	127 **I** Iodine 53	131 **Xe** Xenon 54
133 **Cs** Caesium 55	137 **Ba** Barium 56	139 **La** Lanthanum 57 *	178 **Hf** Hafnium 72	181 **Ta** Tantalum 73	184 **W** Tungsten 74	186 **Re** Rhenium 75	190 **Os** Osmium 76	192 **Ir** Iridium 77	195 **Pt** Platinum 78	197 **Au** Gold 79	201 **Hg** Mercury 80	204 **Tl** Thallium 81	207 **Pb** Lead 82	209 **Bi** Bismuth 83	210 **Po** Polonium 84	210 **At** Astatine 85	222 **Rn** Radon 86
223 **Fr** Francium 87	226 **Ra** Radium 88	227 **Ac** Actinium 89 †															

*58–71 Lanthanum series

140 **Ce** Cerium 58	141 **Pr** Praseodymium 59	144 **Nd** Neodymium 60	147 **Pm** Promethium 61	150 **Sm** Samarium 62	152 **Eu** Europium 63	157 **Gd** Gadolinium 64	159 **Tb** Terbium 65	162 **Dy** Dysprosium 66	165 **Ho** Holmium 67	167 **Er** Erbium 68	169 **Tm** Thulium 69	173 **Yb** Ytterbium 70	175 **Lu** Lutetium 71

†90–103 Actinium series

232 **Th** Thorium 90	231 **Pa** Protactinium 91	238 **U** Uranium 92	237 **Np** Neptunium 93	242 **Pu** Plutonium 94	243 **Am** Americium 95	247 **Cm** Curium 96	245 **Bk** Berkelium 97	251 **Cf** Californium 98	254 **Es** Einsteinium 99	253 **Fm** Fermium 100	256 **Md** Mendelevium 101	254 **No** Nobelium 102	257 **Lr** Lawrencium 103

Key:

a **X** b

a = relative atomic mass
X = atomic symbol
b = atomic number

1 Principles of chemistry 1

Practical work

Example

1 A student investigated what happened when some crushed ice was heated using a Bunsen burner. The diagrams show her investigation at the start and the end.

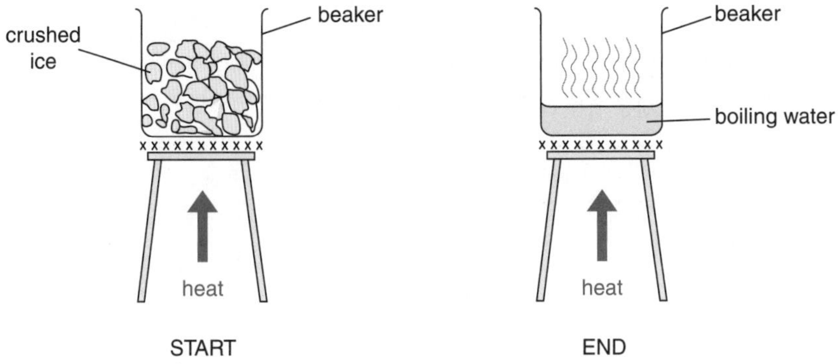

a) At first the solid ice turns to water. Give the name of the process and describe the changes in the arrangement, and movement of the particles as the ice becomes water. *(3)*

Melting → particles heat up - vibrate and move apart bonds less strong

b) After further heating the water starts to boil. After 5 minutes the level of water in the beaker is half the original water level. Explain why the water level has dropped by describing the changes in the movement and energy of the particles as the water's temperature approaches boiling point. *(4)*

As it reaches boiling point it begins to evaporate as water turns to steam due to vibration + movement of particles

c) The water vapour turns back to liquid water on a nearby cold glass. Give the name of the process, and describe how the arrangement and energy of the particles changes as the gas becomes a liquid. *(3)*

condenses → particles lose energy, move more slowly.

(Total for question = 10 marks)

Student 1 response Total 3/10	Marker comments and tips for success
a) *This is <u>melting</u>. ✔ The particles become further away and <u>free to move</u>. ✔*	The process is named correctly, but the description of the arrangement and movement is poor. When you ... ut how the experiment relate...
b) *There is less water in the beaker. The particles have left the beaker and moved into the air. They have more energy.* O	This is a descripti... ...bservations, just repeats them. ... taken place.
c) *Evaporation, as the energy increases the particles get closer together and become water.* ✔	The particles do g... s evaporation, whic... n carefully before answering.

• name process
• particle arrangement
• vibrations / movement
• Amount + explain drop in water level

solid → liquid → gas.
use these words

1 Principles of chemistry 1

Student 2 response Total 7/10	Marker comments and tips for success
a) The process is <u>melting</u>. ✔ In ice the particles are held together in a <u>regular arrangement</u> and are <u>not free to move</u>. <u>They vibrate</u>, getting faster as melting point is reached. ✔	The process is named correctly, and gains a mark. The answer is then split into two parts dealing with the arrangement and movement of particles in ice, and then in water.
As the ice melts the particles become <u>free to move around each other</u> having <u>no regular arrangement</u>, ✔ they are just random.	If you separate your answer into parts like this it is easier for the examiner to find the credit-worthy parts and give you the marks.
b) Some of the water has evaporated after five minutes at boiling point.	The question does not ask for the name of the change of state, so no marks.
This happens because the water particles gain enough energy to <u>separate from each other and go into the air</u>. ✔	The description of the particles' movement gains a mark. An explanation for the drop in water level is needed for a second mark.
As they gain energy they move quicker.	To gain a mark the answer needs to relate the particle energy and movement to temperature. You need to understand the practical described before making the answer.
c) The water appears by <u>condensation</u>. ✔ In the gas the particles are arranged randomly and are spread apart. As the water vapour cools the particles lose energy ✔ and get closer together and form a liquid on the cold surface. ✔	This is a good, well-structured answer. The process is correctly named, with good descriptions.

Practice questions

2 A student used a simple distillation apparatus to obtain some pure water from some muddy water.

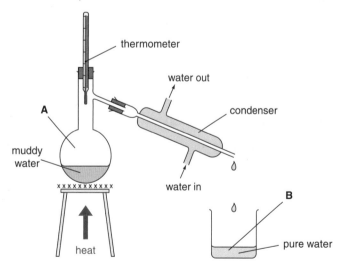

a) Describe the arrangement and movement of the particles at point A. (3)

b) Describe the arrangement and energy of the particles at point B compared to point A. (4)

c) The student left the mud in the bottom of the flask to dry. Describe the movement, arrangement and energy of the solid mud particles. (3)

At point A they begin as a liquid w/ particles in a loose arrangement so can move/slide past each other As the water → gain energy + particles move apart to form gas → volume in beaker decreases

b) At point B the particles of gas have lost their energy forming a liquid. They go from no arrangement to a more structured form where the particles are free to move + slide past each other.

c) As the water evaporates and the mud cools the particles lose energy and move closer together in a regular structure w/ little to no movement between particles.

Practical work

3 A student placed a small purple crystal of potassium manganate(VII) in the bottom of a test tube containing 10 cm³ of water. The diagram shows the test tube immediately after adding the crystal, and one week later.

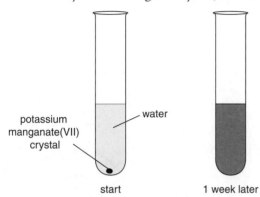

start   1 week later

a) Describe the arrangement of the particles in the crystal of potassium manganate(VII). (2)

It is a solid so the particles are packed closely together in a regular arrangement w/ no movement

b) The student left the test tube for a week.
  i) Describe the appearance of the test tube and contents after one week. (2)

  Still a liquid. The crystal has disappeared / dissolves, solution has turned black.

  ii) Explain what has happened in terms of the water and crystal particles. (3)

  iii) Name the process that has occurred (1)

c) The student then:
  - shook 1 cm³ of the solution in another test tube with 9 cm³ of water
  - removed 1 cm³ of the new solution and placed it in another test tube with 9 cm³ of water
  - repeated the two steps until the solution looked like water.

  Each time the solution was diluted the colour became paler. Suggest how this investigation can be used as evidence that particles have a very small size. (3)

  In the small amount of water the colour is strong as the particles are mixed thoroughly throughout.
  When diluted the particles decrease in concentration forming a paler colour.
  The particles are still present but are so small they are unnoticeable when mixed w/ large amounts of water.

1 Principles of chemistry 1

Using the Periodic Table

You should use a copy of the Periodic Table to help you answer the questions on this section (see page vi).

Example

1 The diagram represents an atom of an element.

  a) i)   State the atomic number and mass number of this element. *an=6 mn=12* (2)
     ii) State the electronic configuration of this element. *(2,4)* (1)
     iii) Name this element. *Carbon* (1)
  b) i)   Which group is this element in? *group 4* (1)
     ii) Name a different element in the same group. *Silicon* (1)
     iii) Name a different element in the same period. *Boron* (1)
     iv) Explain how you chose your answer to part ii). 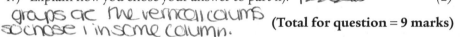 (2)
          groups are the vertical columns so chose 1 in same column. (Total for question = 9 marks) 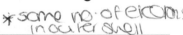 *same no. of e⁻ in outer shell*

Student 1 response	Total 3/9	Marker comments and tips for success
a) i) Atomic number is 12 O, the mass number is 18 O		The student has counted all the particles in the nucleus to obtain an incorrect atomic number, and then added the six electrons to obtain the wrong mass number. Checking the Periodic Table would show that element 12 has a mass of 24, not 18.
ii) 2, 4 ✓		Counting the electrons in each shell gives the electronic configuration.
iii) Magnesium ✓		This uses the incorrect answer from part a) i), but correctly identifies the element in the Periodic Table with atomic number 12, so gains an *error carried forward* mark. This stops the student being penalised twice for the same error.
b) i) Group 4 ✓		The electron diagram shows it is a Group 4 element. This provides an opportunity to correct the answers to a) i) and a) iii), as magnesium is a Group 2 element.
ii) Nitrogen O		The answer confuses group with period and names an element in the same period instead of group, then group rather than period.
iii) Silicon O		
iv) If the element is in the same vertical line they have the same number of electrons. O		Insufficient to gain a mark. You need to explain clearly the connection between the group and the number of electrons in the outer shell.

Student 2 response	Total 9/9	Marker comments and tips for success
a) i) Atomic number is 6, ✓ the mass number is 12 ✓		Knowing that the nucleus contains protons and neutrons, the mass number is found by counting all the dots in the nucleus. As there are six shaded and six unshaded dots, the atomic number must be six.
ii) 2, 4 ✓		The electronic configuration can be determined by counting the electrons in each shell.
iii) Carbon ✓		Carbon is the element on the Periodic Table with atomic number 6.
b) i) Group 4 ✓		The group numbers run across the top of the table.
ii) Silicon ✓		Correct choice from the vertical group.
iii) Oxygen ✓		Correct choice from the horizontal period.
iv) It is in the same group as carbon and elements in the same group ✓ have the same number of outer electrons. ✓		The answer clearly states the relationship between the number of outer electrons and the group number.

Using the Periodic Table

Practice questions

2 The diagram below represents an atom of an element.

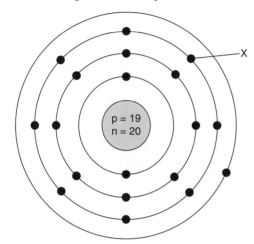

a) i) What is the atomic number of this atom? 19 (1)

ii) What does the n = 20 mean? no. of neutrons = 20 (1)

iii) State the mass number of this atom, and describe how you can calculate it using the diagram. 39 = protons + neutrons (2)

b) i) State the electronic configuration of this atom. (2,8,8,1) (1)

ii) State in which group of the Periodic Table this atom belongs. group 1 (1)

c) i) Name an element with a greater A_r in the same group of the Periodic Table. Rubidium (1)

ii) How many outer electrons will this element have? Explain your answer. (2) 1 = same group = same no. of electrons in outer shell.

3 This question is about atomic structure.

a) Copy and complete this table about the atomic structure of five different atoms from Group 7 of the Periodic Table. Use your copy of the Periodic Table to help you fill in the blanks. (4)

Name	Mass number	Number of			
		Protons	Neutrons	Electrons	Outer electrons
fluorine	19	9	10	9	7
chlorine	35	17	18	17	7
bromine	80	35	45	35	7
iodine	127	53	74	53	7
chlorine	37	17	20	17	7

b) i) State the electronic configuration of fluorine. (2,7) (1)

ii) Astatine is an element in the same group of the Periodic Table. Suggest how many outer electrons it will have. (7) (1)

iii) The table shows a pattern in electronic configurations and group number. What is this pattern? all are in group 7 (1)

c) There are two chlorine atoms in the table, chlorine-35 and chlorine-37.

i) State the difference between the two atoms. chlorine 37 has 2 extra neutrons so is an isotope. (2)

ii) What name is used for different atoms of the same element such as chlorine-35 and chlorine-37? isotopes. (1)

1 Principles of chemistry 1

d) In a sample of chlorine gas a student discovered that there were **three** atoms of chlorine-35 for every **one** atom of chlorine-37. Use this data to calculate that relative atomic mass of chlorine. $35 \times 3 = 105/2 = 52.5$ *(2)*

4 This is an outline of the Periodic Table. The letters below represent elements, but are **not** their chemical symbols.

(Periodic table grid with letters: S, P, T, W, Q, R, X)

a) Which two elements are in the same group? **S R** *(1)*
b) Which two elements are in the same period? **W Q** *(1)*
c) Which two elements have similar chemical properties? **SR / WX** *(1)*
d) Which two elements have the same number of outer electrons? **SR**
   Explain your answer. **SR = same group = same electrons in outer shell** *(2)*
e) Deduce the electronic configurations of elements S, P and T. **(2,2) (2,8,4) (2,8)** *(3)*
f) Name element W. **iron** *(1)*
g) Give the correct chemical symbol for element X. **Au** *(1)*

5 Magnesium and calcium are in the same group of the Periodic Table. Magnesium reacts with many other elements to make compounds.

a) Draw diagrams to show the electronic configuration of:
   i) magnesium *(1)*
   ii) calcium. *(1)*

(Diagrams of Mg and Ca electron shells drawn)

b) Explain in terms of electronic configurations why magnesium and calcium are in the same group of the Periodic Table. **same no. of electrons in outer shell** *(2)*

Argon is in the same period of the table as magnesium. It does not react easily with other elements.

c) Explain in terms of electronic configuration why magnesium is more reactive than argon. *(2)*

d) Draw diagrams to show the outer electronic configurations of:
   i) barium *(1)*
   ii) krypton. *(1)*

(Diagrams of Ba and Kr outer electrons drawn)

Calculations

$$\text{moles} = \frac{\text{Amount in grams}}{\text{Atomic mass}}$$

Example

1. A student dissolved 2.4 g of magnesium ribbon in excess sulfuric acid to make some magnesium sulfate and hydrogen gas.

   $Mg(s) + H_2SO_4(aq) \rightarrow MgSO_4(aq) + H_2(g)$

   a) Calculate the relative formula mass of sulfuric acid. (1) $2 + 16 + 32 = 50$

   b) Calculate the relative formula mass of magnesium sulfate. (1) $12 + 16 + 32 = 60$

   c) Calculate the amount, in moles, of magnesium ribbon used. (2) $= \frac{2.4}{12} = 0.2 \text{ moles}$

   d) How many moles of hydrogen gas were produced? (1)

   e) Calculate the volume, in dm³, of hydrogen gas produced at room temperature and pressure. (2)

   (Total for question = 7 marks)

Student 1 response Total 5/7	Marker comments and tips for success
a) H_2SO_4 $1 + [32 + (16 \times 4)]$ O $= 97$ ✓	The sum is incorrect and gains 0 marks – only one hydrogen has been counted. Check all your calculations twice to make sure you haven't lost a simple mark like this.
b) $MgSO_4$ $24 + [32 + (16 \times 4)]$ ✓ $24 + 96$ $= 120$ ✓	It can be useful in this type of question to work out the formula mass of the sulfate group to save having to work it out every time you need it.
c) Mg used $= \frac{24}{24}$ ✓ $= 0.1$ moles ✓	1 mark for the method and 1 for the calculation. The unit is given in the question, but should be included in the answer.
d) 1 mole O	From the equation, 1 mole of magnesium makes 1 mole of hydrogen gas, but the student forgot that there were only 0.1 moles of magnesium used. The answer should be 1×0.1 mol $= 0.1$ mol.
e) 1×24 ✓ $= 24$ dm³ ✓	The answer gains the mark as the working clearly shows the correct method even though the wrong answer from part d) has been substituted. To prevent the student being penalised twice for the same error the mark is given for the correct method and calculation using the previous wrong answer. This is another example of *error carried forward*.

Student 2 response Total 4/9	Marker comments and tips for success
a) $1 + 1 + 32 + 16 + 16 + 16 + 16$ ✓ $= 98$ ✓	In this answer the four oxygen atoms masses have been added together instead of multiplied, which is an alternative way to get the correct answer.
b) $24 + 98 = 122$ O	This answer loses both marks because of a simple calculation error. SO_4 is 96, H_2SO_4 is 98.
c) $\frac{24}{12}$ O $= 0.2$ moles O	The answer is wrong. The incorrect mass has been used for magnesium.
d) 2 moles O	The student is confused about the subscript numbers. H_2 is one mole of hydrogen gas. You should learn the seven diatomic elements so when you see them in an equation you know that the 1 mole of the gas contains 2 moles of atoms, but only occupies the volume of 1 mole of gas.
e) $0.2 \times 24 = 4.8$ O	The answer uses the wrong number of moles from c) instead of the value from d), so no error carried forward. Also the units have been omitted. You must remember to check that you include the units in your answer with every calculation you do.

1 Principles of chemistry 1

Practice questions

2 A chemical works produces calcium oxide by heating calcium carbonate. This equation represents the reaction:

$CaCO_3(s) \rightarrow CaO(s) + CO_2(g)$

40+12+48 = 100
100 → 56 + 44

a) Calculate the relative formula mass of calcium oxide. (1)

b) Calculate the relative formula mass of calcium carbonate. (1) *44 tonnes*

c) If 100 tonnes of calcium carbonate are converted to calcium oxide each week, calculate the amount, in tonnes, of carbon dioxide gas that is produced. (3)

d) A scientist from the chemical works dissolves 0.88 g of carbon dioxide in 2 dm³ (2000 cm³) of water to form carbonic acid (H_2CO_3).

2+12+48 = 62

i) Calculate the relative formula mass of carbonic acid. (2)

ii) Calculate the amount, in moles, of carbon dioxide used. (2)

moles = mass/mass no = 0.88/62 = 0.0142
0.02m

iii) Calculate the volume of the carbon dioxide, in dm³, that was dissolved in the water at room temperature and pressure. (2)

iv) Calculate the concentration, in mol/dm³, of the carbonic acid. (2) *conc = 0.02/2 = 0.01*

3 Ammonia is an alkaline gas that easily dissolves in water. Ammonia's formula is NH_3. When it dissolves in water it makes ammonium hydroxide solution, NH_4OH.

$NH_3(g) + H_2O(g) \rightarrow NH_4OH(s)$

a) i) Calculate the relative formula mass of ammonia. (2)

ii) Calculate the relative formula mass of ammonium hydroxide. (2)

b) Dissolving 24 dm³ of ammonia gas at room temperature and pressure, in 1 dm³ of water, will make a solution of ammonium hydroxide with a concentration of 1 mol/dm³.

A student dissolved 12 dm³ of ammonia gas in 1 dm³ of water.

i) Calculate the amount, in moles, of ammonia that dissolved. (2)

ii) Calculate the mass, in grams, of the ammonia that dissolved. (2)

iii) What is the concentration, in mol/dm³, of the ammonium hydroxide solution? (1)

c) The ammonium hydroxide solution was then diluted by adding 1 dm³ more water. Calculate the concentration, in mol/dm³, of the diluted solution. (2)

d) Ammonia gas can react with hydrogen chloride gas according to this equation:

$NH_3(g) + HCl(g) \rightarrow NH_4Cl(s)$

i) Calculate the relative formula mass of ammonium chloride. (2)

ii) A student made 13.4 g of ammonium chloride using the reaction. Calculate the amount, in moles, of ammonium chloride that was produced. (2)

Calculations

4 Calcium hydroxide solution can react with carbon dioxide gas to make calcium carbonate.

$Ca(OH)_2(aq) + CO_2(g) \rightarrow CaCO_3(s) + H_2O(l)$

   a) i) Calculate the relative formula mass of water. *(1)*

      ii) Calculate the relative formula mass of calcium hydroxide. *(1)*

      iii) Calculate the relative formula mass of calcium carbonate. *(1)*

   b) 1.48 g of calcium hydroxide is reacted with carbon dioxide.

      i) Calculate the amount, in moles, of calcium hydroxide that reacted. *(2)*

      ii) The reaction produced the same amount, in moles, of calcium carbonate. Calculate the mass, in grams, of calcium carbonate produced. *(2)*

      iii) The reaction required the same amount, in moles, of carbon dioxide gas. Calculate the volume, in dm³, of the carbon dioxide gas that was reacted at room temperature and pressure. *(2)*

5 A student dissolved some magnesium ribbon in excess hydrochloric acid to make some magnesium chloride and hydrogen gas.

$Mg(s) + 2HCl(aq) \rightarrow MgCl_2(aq) + H_2(g)$

   a) Calculate the relative formula mass of hydrochloric acid. *(1)*

   b) Calculate the relative formula mass of magnesium chloride. *(1)*

   c) What is the mass of 1 mole of magnesium? *(1)*

   d) 6 g of magnesium are reacted with excess hydrochloric acid. Calculate the volume of hydrogen gas that would be produced by the reaction. *(2)*

1 Principles of chemistry 2

Charges, chemical formulae and equations

Example

1 A student wanted to find the formula of magnesium oxide.
   The student used this apparatus for the investigation.

   Mass of crucible and lid = 16.81 g

   Mass of crucible, lid and magnesium = 18.24 g

   Mass of crucible, lid and magnesium oxide made = 19.16 g

   a) Calculate the mass, in grams, of magnesium that was used. (1)
   b) Calculate the amount, in moles, of magnesium that was used. (2)
   c) Calculate the mass, in grams, of oxygen that was used. (1)
   d) Calculate the amount, in moles, of oxygen that was used. (2)
   e) Calculate the empirical formula of magnesium oxide. (2)

   (Total for question = 8)

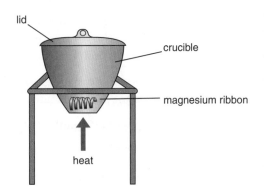

Student 1 response Total 5/8	Marker comments and tips for success
a) $18.24 - 16.81 = 1.43\,g$ ✓	Correct subtraction gains the mark.
b) $\frac{1.43}{24}$ ✓ $= 0.5958333$ O	The calculation is correct, but the answer given is wrong. The units are not necessary as they are given in the question, but you will lose a mark if you give wrong units.
c) $19.16 - 16.81 = 2.35\,g$ O	The wrong subtraction. The mass of oxygen made is the difference in the mass of magnesium and magnesium oxide.
d) $\frac{2.35}{16}$ ✓ $= 0.146875$ ✓	The calculation uses the wrong mass of oxygen, but as this has already been penalised there is an error carried forward mark.
e) $0.5958333 : 0.146875 = ?$ O	This complex ratio needs to be reduced to a simple one. 0.59 is nearly 0.60, and 0.146 is nearly 0.15, which would give a simple ratio of 4 : 1.

Student 2 response Total 8/8	Marker comments and tips for success
a) $18.24 - 16.81 = 1.43\,g$ ✓	Clearly shown calculation with units.
b) $\frac{1.43}{24}$ ✓ $= 0.060$ mol ✓	The calculation is correct and the answer is rounded up to only two significant figures to make later calculations easier.
c) $19.16 - 18.24 = 0.92\,g$ ✓	Correct subtraction gains the mark.
d) $\frac{0.92}{16}$ ✓ $= 0.058$ mol ✓	The calculation is correct. For the unit you can use either mol (the symbol) or moles.
e) $0.060 : 0.058 = 1 : 1$ ✓ so MgO is the formula ✓	Remember that answers from practical work will not always give exact answers. Realising that 0.060 is virtually the same as 0.058 allows the correct formula to be given.

Practice questions

2 a) Iron reacts easily with the oxygen in damp air to form iron(III) oxide.

    i) Give the formula of iron(III) oxide. (1)

    ii) Write a chemical equation for the reaction between iron and oxygen that produces iron(III) oxide. (2)

    iii) State, in terms of electrons, what happens when an oxygen atom becomes an oxide ion. (1)

b) A student heated 14.34 g of an oxide of lead in dry hydrogen. 12.42 g of lead were obtained.

    i) Calculate the mass, in grams, of oxygen that was present in the lead oxide. (1)

    ii) Calculate the amount, in moles, of oxygen that was present in the lead oxide. (2)

    iii) Calculate the amount, in moles, of lead that was used. (2)

    iv) Calculate the empirical formula of the lead oxide. (2)

3 a) A beaker containing some acidified water was electrolysed. Electrolysis can be used to break down compounds like water into the elements they contain. Two gases were collected as shown in the diagram. After 1 hour, 59 cm³ of oxygen and 119 cm³ of hydrogen had been produced at room temperature and pressure.

1 mole of any gas at room temperature and pressure always has a volume of 24 dm³.

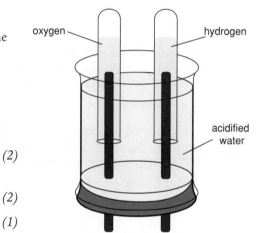

    i) Calculate the amount, in moles, of oxygen produced at room temperature and pressure. (2)

    ii) Calculate the amount, in moles, of hydrogen produced at room temperature and pressure. (2)

    iii) What is the reacting ratio of hydrogen to oxygen? (1)

    iv) What is the empirical formula of water? (1)

b) A compound containing only hydrogen and carbon was found to have a composition of 14.3% hydrogen and a relative molecular mass of 28.

    i) Calculate the amount, in moles, of hydrogen in 100 g of the compound. (2)

    ii) Calculate the amount, in moles, of carbon in 100 g of the compound. (3)

    iii) Calculate the empirical formula of the compound. (2)

    iv) Calculate the molecular formula of the compound. (2)

4 a) A student analysed 31.9 g of anhydrous copper sulfate and found that it contained 12.7 g of copper and 6.4 g of sulfur.

    i) Calculate the mass, in grams, of oxygen the copper sulfate contained. (1)

    ii) Calculate the amount, in moles, of oxygen the copper sulfate contained. (2)

    iii) Use the information in the question to calculate the empirical formula of the copper sulfate. (4)

b) The same student made a sample of hydrated manganese sulfate. The sample weighed 22.3 g. The student then dehydrated the sample, and found the mass was now 15.1 g. Calculate the value of x in the formula $MnSO_4 \cdot xH_2O$. (4)

1 Principles of chemistry 2

Using electronic configurations

Example

1 Chlorine is an element. Chlorine reacts with many other elements. This question is about the compounds that chlorine can make.

   a) i) Chlorine has 17 electrons. Use the Periodic Table to help you to complete a copy of the diagram to show the arrangement of electrons in a chlorine atom. (2)

   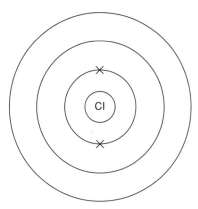

   ii) Chlorine atoms join together to form chlorine molecules, Cl_2. Draw a dot and cross diagram to show how two chlorine atoms join together to form a chlorine molecule. You should only show the outer electrons of each atom. (2)

   b) i) Chlorine can also join together with other atoms such as sodium. Use the Periodic Table to help you to complete a copy of the diagram to show the arrangement of electrons in a sodium atom. (2)

   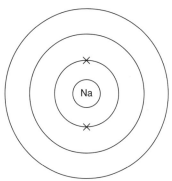

   ii) When chlorine bonds with sodium it forms the compound sodium chloride (NaCl). Draw a dot and cross diagram to explain how the sodium and chlorine atoms are bonded together. You should only show the outer electrons of each atom. (2)

   (Total for question = 8)

Using electronic configurations

Student 1 response	Total 8/8	Marker comments and tips for success
a) i)	[diagram: Cl atom with electron shells 2,8,7 shown as crosses]	The electronic configuration is correct. To work this out, place the first two electrons in the inner shell, then the next eight in the second shell, leaving seven for the outer shell. Check your answer by counting the crosses – there should be 17.
ii)	[diagram: Cl–Cl dot and cross showing shared pair between two Cl atoms, each with surrounding outer electrons]	The number of outer electrons comes from the electronic configuration diagram in part a) i). This diagram clearly shows the two atoms interacting in the molecule of chlorine. Make sure the shared electrons are clearly shown on the same shell.
b) i)	[diagram: Na atom with electron shells 2,8,1 shown as crosses]	This is the correct electronic configuration. You need to know that the first two electrons fill the inner shell, then eight electrons go in both the second and third shells. The fourth shell takes the rest, if the element has 20 electrons or fewer.
ii)	[diagram: Na × → Cl (with dots), then Na⁺ ----- ×Cl⁻ labelled "electrostatic attraction"]	The diagrams clearly show the transfer of the electron and the resulting charges and attraction between the two ions. Adding a label, as here, is a good way to explain your diagrams.

Student 2 response	Total 2/8	Marker comments and tips for success
a) i)	[diagram: Cl atom with three shells, crosses distributed]	1 mark for each correct shell.
ii)	Cl – Cl	The question asks for a dot and cross diagram showing outer electrons. This diagram is a displayed formula. Read the question carefully.
b) i)	[diagram: Na atom with electrons incorrectly placed – 2 in inner shell and 7 in second shell]	The first two electrons have been added to the already filled inner shell and seven have been added to the second shell. This has used up all the electrons, but leaves the third shell with none. Make sure you only put the correct maximum number in each shell.
ii)	[diagram: Na and Cl shown with overlapping circles as if covalent bond]	With the error in b) i) it is now possible to make a covalent bond from the wrong electronic configuration. If you know it should be gain and loss of electrons and it doesn't work, go back and find out where you have gone wrong.

13

1 Principles of chemistry 2

Practice questions

2 a) Carbon has **six** electrons. Copy the diagram and use the Periodic Table to help you to complete it to show the arrangement of electrons in a carbon atom. (2)

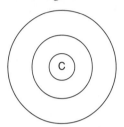

b) Hydrogen has **one** electron. The electronic configuration of hydrogen is shown. Hydrogen and carbon make the molecule methane (CH_4). Draw a dot and cross diagram to show the arrangement of electrons in a molecule of methane. You should only show the outer electrons of each atom. (1)

c) Carbon can also make the molecule carbon dioxide (CO_2).

   i) Oxygen has **eight** electrons. Copy the diagram and use the Periodic Table to help you to complete it to show the arrangement of electrons in a oxygen atom. (2)

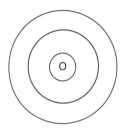

   ii) Draw a dot and cross diagram to show the arrangement of electrons in a molecule of carbon dioxide. You should only show the outer electrons of each atom. (2)

3 a) The electronic configuration of calcium can be written as 2, 8, 8, 2. Use the Periodic Table to write the electronic configurations for the following elements.

   i) sodium (1)
   ii) calcium (1)
   iii) fluorine (1)
   iv) sulfur (1)
   v) hydrogen (1)

b) Draw dot and cross diagrams to show the bonding in the following compounds. You should only show the outer electrons of each atom.

   i) hydrogen sulfide (2)
   ii) sodium fluoride (2)
   iii) calcium fluoride (2)

4 Copy and complete the following chart about the electronic configuration of some ions and atoms. Use the Periodic Table to help you. (5)

Element	Proton number	Electronic configuration of an atom	Electronic configuration of an ion	Number of outer electrons
neon		2, 8	no ions	
beryllium				2
nitrogen	7			
aluminium			2, 8	
phosphorus				5

5 a) Explain using dot and cross diagrams the formation of the following covalent compounds. You should only show the outer electrons of each atom.

   i) hydrogen chloride (HCl) (1)
   ii) ammonia (NH_3) (2)
   iii) water (H_2O) (2)
   iv) nitrogen (N_2) (1)

b) Explain using dot and cross diagrams the formation of the following ionic compounds. You should only show the outer electrons of each atom.

   i) lithium fluoride (2)
   ii) magnesium oxide (2)
   iii) calcium chloride (2)

1 Principles of chemistry 2

Data analysis

Example

1 The table shows the properties of five different substances.

Substance	Melting point / °C	Boiling point / °C	Conducts electricity as a Solid	Conducts electricity as a Liquid
P	−114	−85	no	no
Q	714	1418	no	yes
R	3550	4827	no	no
S	−39	357	yes	yes
T	6	15	no	no

Use the information in the table to answer the following questions.

a) Which substance has the lowest melting point? (1)

b) Which substance is a liquid at room temperature (20 °C)? (1)

c) Which substance is a metal? Explain your answer. (2)

d) Which substance has a giant covalent structure? Explain your answer. (2)

(Total for question = 6 marks)

Student 1 response Total 1/6	Marker comments and tips for success
a) S O	The melting point of S (−39 °C) is the smallest negative number, but remember that with negative numbers, the larger the number the lower the value.
b) T O	This is a gas, not a liquid. For a substance to be a gas at room temperature of 20 °C, the boiling point must be more than 20 °C and the melting point less. Read the data very carefully.
c) Q, as it conducts electricity. O	In a question like this every column of data has a purpose. You need to look at two pieces of data, not one. To be a metal the substance must conduct as both a solid and as a liquid.
d) R ✓, as it cannot conduct electricity.	1 mark for the correct substance. For the second mark you need to explain using two sets of data. Read the table headings carefully and decide which data to use. It can help to read the questions and then go back to the table. Use what you know about key terms like *ionic* and *covalent* to help understand the data.

Student 2 response Total 6/6	Marker comments and tips for success
a) P ✓	Correctly identifies −114 °C as the lowest melting point.
b) S ✓	Correct choice.
c) S ✓, as metals conduct electricity as both solids and liquid. ✓	You do not have to memorise data, but you must know the key features of each type of substance and be able to find it in the table, e.g. metals conduct electricity as both solids and liquids, and are usually solids at room temperature.
d) R, ✓ as it has high melting and boiling points, and doesn't conduct electricity, ✓ so it must be covalent with a crystal lattice.	A clear answer which uses information from more than one column of the table to choose the substance and explains the choice using known properties of covalent substances.

Practice questions

2 The table shows some information about two substances, chlorine and sodium chloride. Use information from the table to help you answer these questions.

Property	Chlorine	Sodium chloride
melting point / °C	−101	808
boiling point / °C	−34	1465
solubility in water / grams/litre	7	315
electrical conductivity as a liquid	poor	good
electrical conductivity when dissolved in water	poor	good
structure at room temperature (20 °C)		

a) Which of the two substances has ionic bonding? Give **two** pieces of information from the table to support your answer. (3)

b) Explain, using information from the table, the differences in solubility of sodium chloride and chlorine. (2)

c) Explain, using information from the diagrams, why sodium chloride has a high melting point and chlorine has a low melting point. (4)

3 The table shows the formulae, boiling points and bonding in the chlorides of some elements.

Atomic number of element	3	4	5	6	7	8	9	10
Formula of chloride	LiCl	$BeCl_2$	BCl_3	CCl_4		OCl_2	none	no chloride
Boiling point of chloride / °C	1382	547	13	76	71	2	–	–
Bonding	ionic			covalent	covalent	covalent	–	–

a) The element with atomic number 7 has symbol N. Suggest the formula of its chloride. (1)

b) Suggest the type of bonding in BCl_3. Explain your answer. (2)

c) Describe the trend in boiling points of the chlorides. (2)

d) State the atomic numbers of the metallic elements. (1)

e) Explain why the element with atomic number 10 has no chlorides. (2)

f) If chlorine makes an ion with a charge of −1, what is the charge on the ion of element with atomic number 3? Explain your answer. (2)

1 Principles of chemistry 2

4 Carbon is an element that can exist in several different forms. The table gives some information about two of these forms, diamond and graphite.

Property	Diamond	Graphite
melting point / °C	3550	3727
number of C–C bonds per atom	4	3
electrical conductivity	poor	good
uses	used to cut very hard substances	used as a high temperature lubricant
structure		

a) Use information in the table to suggest why diamond and graphite both have high melting points. *(3)*

b) Use information from the table to explain why diamond is used to cut glass. *(2)*

c) Use information in the table to explain why graphite can be used as a high temperature lubricant. *(3)*

d) Use information in the table to suggest why diamond does not conduct electricity, but graphite does. *(4)*

2 Chemistry of the elements

Using the Periodic Table

Example

1 This is an outline of the Periodic Table. The letters below represent elements, but are **not** their chemical symbols.

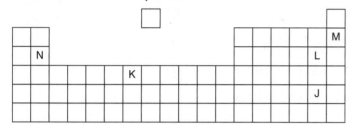

a) Which elements are in the same period? (1)

b) Which element is a transition metal? (1)

c) What is the charge on an ion of element L? (1)

d) Which element is very unreactive? Explain why. (2)

(Total for question = 5 marks)

Student 1 response Total 4/5	Marker comments and tips for success
a) L and N ✔	Clearly identifies elements in the same period (row).
b) M ✗	M is a noble gas; K is the transition metal. Remember that the transition metals are in the middle block of the table.
c) negative ✔	Negative gains the mark, but you should mention how many negative charges, as some elements have two negative charges.
d) M ✔ It has no outer electrons to use to react. ✔	This gains both marks, but a better explanation would refer to a full outer shell of electrons instead of no outer electrons.

Student 2 response Total 4/5	Marker comments and tips for success
a) L and J ✗	The answer confuses periods (rows) with groups (columns)
b) K ✔	Correctly locates the transition metals in the middle block of the Periodic Table.
c) – ✔	The mark is gained here as there is only one minus charge.
d) M ✔ It has a full outer shell of electrons ✔ to react with.	1 mark for identifying the unreactive noble gas. The explanation is poor, but gains the mark. Noble gases do not react as they have full outer shells.

Practice questions

2 This table shows the electronic configurations of several elements.

Element	Electronic configuration
A	2, 8
B	2, 8, 7
C	2, 8, 1
D	2, 6
E	2, 1

a) State an element in the table which will be a non-metal. (1)

b) Which element in the table will be very unreactive? (1)

c) State the element that will have an ion with a 2⁻ charge. (1)

d) Which **two** elements will have similar chemical reactions with water? (1)

e) Element F is directly beneath element D in the Periodic Table. Write its electronic configuration. (1)

f) Use the Periodic Table to help you name element D. (1)

3 Use information from the diagrams below and the Periodic Table to help you answer this question.

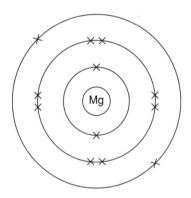 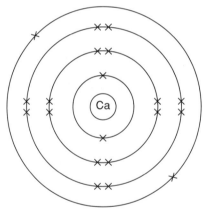 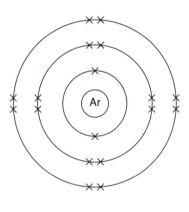

a) Name an element with two more outer electrons than magnesium. (1)

b) Name an element that has the same number of outer electrons as calcium, and is heavier. (1)

c) Name the element that has three fewer electrons than argon. (1)

d) Explain why argon is very unreactive. (1)

e) A calcium ion has the same electronic configuration as an argon atom. Explain why. (2)

4 The table shows the formulae, and pH of the aqueous solution of the oxides of elements in the same period of the Periodic Table.

Element	sodium	magnesium	aluminium	silicon	phosphorus	sulfur	chlorine	argon
Formula of oxide	Na_2O	MgO	Al_2O_3	SiO_2	P_4O_6	SO_2	Cl_2O	no oxide
pH of aqueous solution	14	11	insoluble	insoluble	2	1	over 3	no oxide

a) Describe how the pH of the aqueous solution of oxides changes across the period. (3)

b) What does the table show about the oxides of elements in the middle of the period? (1)

c) Use information in the table to suggest a method of recognising a non-metal element using its oxide. (2)

d) Suggest the likely pH of the aqueous solutions of the oxides of these elements:

   i) nitrogen (1)

   ii) bromine (1)

   iii) lithium (1)

   iv) germanium. (1)

5 Sodium oxide forms an alkaline solution when dissolved in water. Sulfur oxide forms an acidic solution when dissolved in water.

a) i) What type of element normally has alkaline oxides? (1)

   ii) Use the Periodic Table to name a different element that has an alkaline oxide. (1)

   iii) What type of element normally has acidic oxides? (1)

   iv) Use the Periodic Table to name a different element that has an acidic oxide. (1)

b) i) What type of element normally conducts electricity? (1)

   ii) Use the Periodic Table to name a different element that conducts electricity. (1)

2 Chemistry of the elements

Using electronic configurations

Example

1. Sodium is a metal in Group 1 of the Periodic Table. Sodium's atomic number is 11. Sodium reacts violently with water.

   a) What is the electronic configuration of a sodium atom? (1)

   b) Describe what you would observe during the reaction of sodium with water. (4)

   c) Balance the chemical equation for the reaction of sodium with water. (1)

   __ Na(s) + __ $H_2O(l)$ → __ NaOH(aq) + __ $H_2(g)$

   d) Explain, in terms of the arrangement of electrons in the atoms, why the reactivity of the metals with water increases down Group 1. (3)

   (Total for question = 9 marks)

Student 1 response Total 3/9	Marker comments and tips for success
a) 2, 8 O	This is the electronic configuration of the ion, not the atom.
b) It <u>floats</u> ✓ and <u>fizzes</u> ✓ around. The water turns blue, and the lump eventually <u>disappears</u>.	Two points gain 2 marks. The answer then describes the water turning blue, remembered from a demonstration as the sodium hydroxide made reacted with universal indicator. This is correct, but the question does not mention an indicator. No mark for 'disappear' – the sodium reacts.
c) $3Na(s) + 3H_2O(l) \rightarrow 3NaOH(aq) + 3H_2(g)$ O	This is not balanced. There are only six hydrogen atoms as reactants but nine hydrogen atoms as products.
d) There are more electrons. For instance lithium has three electrons and sodium 11. This means that they are easier to lose as they are <u>further away from the nucleus</u> ✓ as there are more electrons.	The question wanted electronic arrangements, not number of electrons. There is 1 mark for the increasing distance of the electrons from the nucleus. The other 2 marks are for explaining how the attraction to the positive nucleus lessens owing to more shells, which makes the outer electron easier to lose.

Student 2 response Total 7/9	Marker comments and tips for success
a) 2, 8, 1 ✓	You can either draw the electron shells, or just give the correct numbers to gain the mark. Drawing the shells means you can count up the electrons at the end to check you have 11.
b) The sodium <u>floats</u> ✓ and <u>turns into a ball</u> ✓. It moves around getting smaller as it <u>reacts</u> ✓ with the water. Eventually it disappears.	Good description worth 3 marks. Mention of the evidence for a gas would gain the final mark. The word 'disappears' is used, but in this case it is associated with getting smaller through reaction.
c) $4Na(s) + 4H_2O(l) \rightarrow 4NaOH(aq) + 3H_2(g)$ O	The balancing is tricky here. Always start with the metal, then the other elements, leaving oxygen and hydrogen to be last. It should be $2H_2$.
d) As you go down the group there is a fresh electron shell, e.g. Na has 2, 8, 1, and K has 2, 8, 8, 1. This extra shell <u>increases the distance from the outer electron to the nucleus</u> ✓, making the outer electron <u>less attracted to the nucleus</u> ✓, so the K electron is <u>lost more easily</u> ✓ than the Na.	The answer clearly refers to the electron arrangements with two examples. Using the symbol rather than the name for the elements saves time in making the answer. The answer clearly refers to the effect of more shells, and the lessening attraction of the nucleus on the outer electron.

Using electronic configurations

Practice questions

2 The table below shows the atomic number, electronic configuration, and the relative distance of the outer electron from the nucleus of some elements.

Group 1 element	Atomic number	Electronic configuration
lithium	3	
sodium	11	2, 8, 1
potassium	19	2, 8, 8, 1

a) What is the electronic configuration of lithium? (1)

b) Describe the relative reactivities of the three elements. (2)

c) Describe the trend of the relative distance of the outer electron to the nucleus in comparison to the atomic number. (1)

d) Rubidium is in the same group of the Periodic Table. Suggest its reactivity in comparison to potassium. (1)

e) Explain, in terms of the arrangement of electrons in the atoms, the differences in reactivity down the group of elements. (3)

3 The diagram shows the electronic configuration of a chlorine atom.

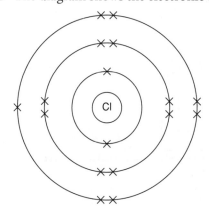

a) Draw a similar diagram to show the electronic arrangement of fluorine. (2)

b) Bromine and iodine are elements in the same group as fluorine and chlorine.

  i) State the group number of this group. (1)

  ii) State the number of outer electrons in a bromine atom. (1)

c) Fluorine is a more reactive element than chlorine. Suggest which of the elements in each pair will be most reactive.

  i) chlorine and bromine (1)

  ii) bromine and iodine (1)

  iii) Explain your answers to parts i) and ii) in terms of the reactivity of Group 7. (1)

4 When chlorine gas is bubbled through sodium bromide solution the colourless solution turns orange.

a) Why does the solution turn orange? (1)

b) Complete the ionic equation for the reaction. (2)

$Cl_2(g) + \underline{\quad} Br^-(aq) \rightarrow \underline{\quad}(aq) + \underline{\quad}(aq)$

2 Chemistry of the elements

  c) What is the electronic configuration of:

  i) a chlorine atom (1)

  ii) a chloride ion? (1)

  d) Explain in terms of electrons how the chlorine atoms become chloride ions during the reaction. (2)

5 A student dissolved some hydrogen chloride gas in water to make some hydrochloric acid. The student knew that the hydrogen chloride gas could not conduct electricity, and was surprised to find that when dissolved in water it now conducted electricity.

  a) Draw a dot and cross diagram to show the arrangement of electrons in hydrogen chloride. You should show only the outer electrons of each atom. (2)

  b) Complete the equation to show what happens when hydrogen chloride dissolves in water. (2)

  $HCl(g) \rightarrow$ _____ (aq) + _____ (aq)

  c) Explain in terms of the arrangement of electrons what happens to:

  i) the hydrogen atom in hydrogen chloride when it dissolves (1)

  ii) the chlorine atom in hydrogen chloride when it dissolves. (1)

  d) Explain why the solution of hydrochloric acid can conduct electricity but liquid hydrogen chloride cannot. (3)

6 a) Copy and complete the table about Group 1 and 7 elements. (5)

Element	Appearance at room temperature	Electronic configuration of the atom	Electronic configuration of the ion.
lithium	dull grey solid	2, 1	
fluorine	colourless gas		2, 8
sodium	light grey solid		2, 8
chlorine		2, 8, 7	
potassium	dark grey solid	2, 8, 8, 1	

  b) Explain why the electronic configuration of a sodium ion is the same as that of a fluoride ion. (2)

  c) Explain, in terms of the arrangement of electrons in the atoms, why potassium is more reactive than lithium. (3)

Charges, chemical formulae and equations

Example

1 A student used the apparatus below to investigate the percentage of oxygen in air. The copper was heated and reacted with the oxygen in the air to make copper oxide.

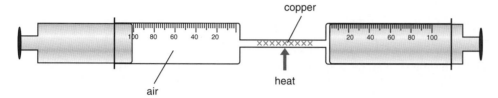

The 100 cm³ of air in the apparatus was passed repeatedly over the hot copper until no further loss in volume occurred. The apparatus was allowed to cool and the volume of air remaining was found to be 79 cm³.

a) Write a chemical equation for the reaction between copper and oxygen. (2)

b) Calculate the percentage of oxygen in the air. (2)

c) The copper has reacted with the oxygen. What name is given to the type of change when a substance gains oxygen? (1)

d) State in terms of electrons what happens when a copper atom becomes a copper ion. (1)

(Total for question = 6 marks)

Student 1 response Total 1/6	Marker comments and tips for success
a) $Cu + O \rightarrow CuO$ O	The formulae for Cu and CuO are correct, but O should be O_2. You must remember the elements that have diatomic molecules.
b) 79% O	The question has not been read properly. The 79 cm³ is the volume of gas at the end after the oxygen has been removed. This loses both marks, as there is no indication that the student knows how to calculate a percentage. Always show your working.
c) oxidation ✔	You need to remember that oxidation is when a substance gains oxygen.
d) The copper loses electrons. O	This is what happens, but the answer doesn't relate the charge on a copper ion (Cu^{2+}) to the *number* of electrons lost. You should remember the charges on copper, zinc and iron, as these are hard to work out from the Periodic Table.

Student 2 response Total 6/6	Marker comments and tips for success
a) $2Cu(s) + O_2(g) \rightarrow 2CuO(s)$ ✔✔	Both formulae and balancing are correct, and gain the marks. The answer includes state symbols, but these are not needed unless asked for in the question.
b) $100 - 79 = 21$ ✔ $\frac{21}{100} \times 100 = 21\%$ ✔	The answer shows a clear understanding of how to calculate the volume of oxygen and then the percentage in the air.
c) oxidisation ✔	Make sure you spell technical words correctly. On this occasion it doesn't matter, but for very similar technical words you would lose the mark for incorrect spelling.
d) The copper atom has <u>lost two electrons</u> ✔ to become an ion.	A clear description of what has happened in terms of electrons. You could also describe this using a half-equation: $Cu^{2+} + 2e^- \rightarrow Cu$

2 Chemistry of the elements

Practice questions

2 Magnesium powder was used in early photography to provide flash lighting. When burnt the magnesium powder reacted with the oxygen in the air producing large amounts of bright white light.

   a) Complete the chemical equation for the burning of magnesium in air. (2)

   ___ Mg(s) + O_2(g) → _____ .

   b) The magnesium has reacted with the oxygen. What name is given to the type of change when a substance gains oxygen? (1)

   c) State in terms of electrons what happens to an oxygen atom when it becomes an oxygen ion. (1)

   d) Magnesium can also burn in carbon dioxide gas, producing dense black smoke.

   ___ Mg(s) + _____ → _____ + C(s)

   i) Complete the chemical equation for the burning of magnesium in carbon dioxide. (2)

   ii) Which substance will be the dense black smoke? (1)

   iii) The carbon dioxide has reacted with the magnesium and lost oxygen. What is the name given to the type of change when a substance loses oxygen? (1)

3 Sulfur is present in diesel fuels. When the fuel is burnt inside a diesel engine the sulfur reacts with oxygen to form sulfur dioxide (SO_2) gas.

   a) Use the Periodic Table to find the charge on an oxygen ion. (1)

   b) Write a chemical equation for the reaction of sulfur with oxygen. (2)

   c) The oxygen has reacted with the sulfur. What is the name given to the type of change when a substance gains oxygen? (1)

   d) Explain why sulfur dioxide is an atmospheric pollutant. (2)

4 A student was interested in finding the percentage of oxygen in air. The student placed some iron wool in the bottom of a glass measuring cylinder and then inverted it over water as shown in the diagram. After a week the water level inside the cylinder was at the 40 cm³ mark.

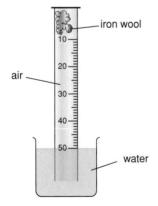

   a) Copy and complete the equation to show how the iron reacted with the air. (2)

   _____ + _____ → ___ Fe_2O_3(s)

b) What is the charge on the iron ion in the iron oxide in the equation? (1)

c) Calculate the percentage of oxygen in the air. (2)

d) The iron has reacted with the oxygen. What is the name given to the type of change when a substance gains oxygen? (1)

e) There is a second type of iron oxide with the formula FeO. What is the charge of the iron ion in FeO? (1)

f) State, in terms of electrons, what happens to the iron atom in FeO when it becomes an iron ion. (1)

5 Zinc carbonate ($ZnCO_3$) can be converted to zinc oxide by strongly heating the zinc carbonate.

a) Name the gas that is released when zinc carbonate is converted to zinc oxide. (1)

b) Explain how the gas released may contribute to climate change. (2)

c) Copy and complete the equation to show the conversion of zinc carbonate to zinc oxide. (2)

$ZnCO_3(s) \rightarrow$ _____ + _____

d) The oxygen in zinc oxide can be removed if the zinc oxide is heated strongly with carbon.

   i) Name the gas that is produced. (1)

   ii) The carbon gains oxygen so it is oxidised. What is the name given to the type of change when a substance loses oxygen? (1)

   iii) What is the name we give to reactions when one substance gains oxygen, and another substance loses oxygen? (1)

6 a) Copy and complete the chart below to show the composition of air. (3)

Gas	Percentage in air
	21
nitrogen	79
carbon dioxide	
	0.93
water vapour	variable

b) Give **two** reasons why the percentage of water vapour in the air is variable. (2)

c) Magnesium is easily oxidised to magnesium oxide.

   i) Write a chemical equation for the reaction. (2)

   ii) Name the substance that is oxidised. (1)

2 Chemistry of the elements

Practical work 1

Example

1 A student visits his local graveyard. He sees several headstones made from different materials.

He notices that the older gravestones have been corroded by acid rain. It appears to be worse for tombstones made from marble (calcium carbonate).

The student thinks the marble corrodes faster in acid rain than other types of stone.

a) Describe an experiment that the student could do to test his idea. (5)

b) Acid rain reacts with calcium carbonate to produce carbon dioxide.

Describe a test to identify that the gas produced is carbon dioxide. (2)

(Total for question = 7 marks)

Student 1 response Total 7/7	Marker comments and tips for success
a) Collect five or six different types of stone used in headstones, such as granite, sandstone, marble, limestone. Weigh each sample ✔ and place in a beaker containing 50 cm³ of acid rain. ✔ This could be quite dilute sulfuric acid to act as the acid so that the reaction will be faster than happens in graveyards so we could get results quite soon. Put a second piece of marble in beaker containing pure water as a control. ✔ Leave the beakers for a few days, such as a week. ✔ Remove and wash the stones, dry them and re-weigh them. ✔ Find the percentage mass lost by each stone.	This answer clearly describes how to carry out a test that will produce reliable results. The control variables are all correctly identified; a sensible method of speeding up the experiment is given that controls the concentration of the acid rain. There is a clear strategy for making the results for each stone comparable.
b) Collect and shake the gas with some limewater. ✔ If the gas is carbon dioxide, the limewater will produce a white precipitate. ✔	Good description. The answer also uses the correct scientific word, 'precipitate' for the white 'cloudiness' that appears in the solution.

Student 2 response Total 2/7	Marker comments and tips for success
a) Put a different stone in several test tubes. Cover each stone with some acid rain. ✔ Leave them for a week, then look at the stones to see which has corroded most. ✔ To make it a fair test you could keep the tubes in the same place in your lab.	This is a poor method that gives insufficient explanation. You need to give a description of how to control the key variables of mass/size of stone, not the volume of acid rain. Looking to see which has corroded most will not give results that can be compared, as there is no attempt to either explain how you would tell which has corroded most or measure it. Wherever possible you should devise a method of measuring the change that allows for a comparison. In this case the easiest is to use the change in mass. The fair test statement is really about control measures, but is not in enough detail to gain any marks.
b) Use limewater	When describing how to do a test, you should mention how to do it as well as the chemical(s) needed, with the positive result you would expect from the test.

Practice questions

2 A student wanted to find out how easily different metal carbonates decomposed on heating.

He placed a sample of a metal carbonate into a test tube and heated it. The carbon dioxide given off was collected in a measuring cylinder as shown.

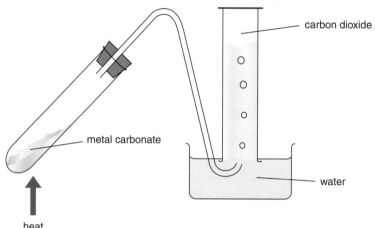

He heated three other metal carbonates in turn and measured the volume of carbon dioxide gas produced by 3 minutes of heating.

The results are given in the table.

Metal carbonate	Volume of gas produced after heating for 3 minutes / cm³
sodium carbonate	5
zinc carbonate	120
copper carbonate	165
magnesium carbonate	75

a) Use the results to identify, with a reason, which metal carbonate decomposed most easily. (2)

b) State **two** things that the student must do to make sure the experiment is valid. (2)

c) The student was warned about the danger of 'sucking back'. Explain what this is, and how to avoid it. (2)

3 Marsha has been told by a friend that carbon dioxide is used as a fire extinguisher because it is heavier than air. This means that the carbon dioxide can cover the fire and prevent oxygen reaching the fire.

Marsha always thought carbon dioxide was lighter than air.

a) Describe an experiment that Marsha's friend could do to prove to Marsha that carbon dioxide is heavier than air. (4)

b) Write a chemical equation to show how carbon dioxide can be made by the reaction of calcium carbonate ($CaCO_3$) and dilute hydrochloric acid (HCl(aq)). (2)

4 A student visited a museum and looked at some old metal sculptures. The student noticed that some sculptures were more corroded than others. She thought this was because some metals are more reactive to acids in rainwater than others.

   a) Describe an experiment that the student could do to test her idea. (5)

   b) Describe a physical test the student could do to show that rain water is not pure water. (2)

5 **a)** Copy and complete the table below about the reactions of some elements with oxygen in the air.

    The first line has been completed. (6)

Element	Observations during reaction	pH of aqueous solution of oxide	Is the oxide an acid or a base?
calcium	burns with a red glow	13	base
carbon			
magnesium			
sulfur			

   b) Describe how the results in the table can be used to decide if an element is a metal or a non-metal. (2)

   c) Describe a test you could use to show that the solution of calcium oxide was made with water as the solvent and not a different solvent. (2)

   d) Describe how you could test a liquid to show it was pure water. (2)

Practical work 2

Example

1. Gypsum is used to make plaster for house walls. The ions present in gypsum can be identified using chemical tests.

   A student carried out a flame test on some gypsum. It burned brick red.

   a) Name the positive ion that is present in gypsum. *(1)*

   The student then dissolved some gypsum in water and tested the solution to see if it contained chloride ions.

   b) Describe how you would test for the presence of chloride ions. *(3)*

   c) The test for chloride ions did not work. The student then tested the gypsum for sulfate ions and obtained a positive result. Describe how the student tested for sulfate ions. *(3)*

   d) Write the chemical formula for gypsum. *(1)*

   (Total for question = 8 marks)

Student 1 response — Total 6/8	Marker comments and tips for success
a) calcium ✔	Correct flame test colour for calcium.
b) Add <u>silver nitrate solution</u>. ✔ A chloride will produce a <u>white cloud</u> in the water. ✔	The addition of dilute nitric acid has been forgotten, so 1 mark is lost. With the tests for halogen ions you always add dilute nitric acid to make sure there are plenty of nitrate ions to react. There is 1 mark for the silver nitrate and 1 mark for the white 'cloud', although solid or precipitate would be much better descriptions. Remember to use the correct scientific words.
c) Add <u>barium chloride solution</u>. ✔ A sulfate will produce a <u>white cloud</u> in the water. ✔	1 mark is lost for forgetting to mention adding dilute hydrochloric acid. With barium chloride you need to add dilute hydrochloric acid to ensure enough chloride ions for a positive result. Remember that you add the same acid that makes the salt of the chemical you are adding to the unknown solution.
d) $CaSO_4$ ✔	The two test results have been used to correctly identify the ions and the formula is correct.

Student 2 response — Total 1/8	Marker comments and tips for success
a) lithium O	Lithium is deep crimson, not brick red. You need to learn the flame test colours for Li^+, Na^+, K^+ and Ca^{2+}.
b) Add some dilute hydrochloric acid. It will fizz. O	This is a poor description of the test for carbonates. The question wanted the test for a chloride. You must learn the tests for sulfate, carbonate and halide ions. To get 3 marks you need to give both correct chemicals and the positive result.
c) Add <u>silver nitrate solution</u>. It will produce a <u>white suspension</u>. ✔	For practicals involving identification of ions and gases you have to know the tests. Here the solution to be added is wrong, but the description of the positive result gains a mark.
d) $LiSO_4$	This formula is wrong. Writing Li_2SO_4 would have gained the mark under the error carried forward rule. Always check the charges on ions so you get the correct formula.

2 Chemistry of the elements

Practice questions

2. Ammonium carbonate ((NH_4)$_2CO_3$) is sometimes called *baker's ammonia*. It is used as a raising agent in bread and cake, and is present in some baking powders.

   a) Describe how you would test a solution of baker's ammonia for the presence of carbonate ions. (3)

   b) Describe how you would test a solution of baker's ammonia for the presence of ammonium ions. (3)

   c) Use the Periodic Table to help you calculate the relative formula mass of baker's ammonia. (2)

3. Bordeaux mixture is used as a fungicide. It is made from a mixture of calcium hydroxide ($Ca(OH)_2$) and copper sulfate ($CuSO_4$).

   a) Describe how you would test a solution of Bordeaux mixture for the presence of copper ions. (2)

   b) Describe how you would test a solution of Bordeaux mixture for the presence of sulfate ions. (3)

   c) Copper ions give a green flame test result. Suggest why it is not possible to show the presence of calcium ions in Bordeaux mixture using a flame test. (2)

4. Potassium bromide (KBr) is used in some medicines. To show the presence of potassium bromide in a medicine you have to test for potassium and bromide ions.

   a) Describe how you would test a solution of a medicine for the presence of bromide ions. (3)

   b) Describe how you would test a solution of a medicine for the presence of potassium ions. (3)

   c) You can use the same test for iodide ions as for bromide ions. What would you expect the result to be for iodide ions? (1)

5. A chemical factory that uses hydrogen, oxygen, ammonia and chlorine gases has been badly damaged by a falling tree. The room where the gases are stored has filled with gas from a leaky gas cylinder. The emergency response team has asked you to identify which gas has escaped.

   a) Describe how you would test the gas to decide if it was chlorine or ammonia. (3)

   b) Describe how you would test the gas to show it was oxygen. (1)

   c) Describe how you would test the gas to show it was hydrogen. (1)

6. A teacher in a laboratory has made up some iron(II) sulfate solution for use in the laboratory later on in the week. When the solution was used, the experiment failed to work. One of his colleagues looked at the solution and told him that the solution contained iron(III) ions, not iron(II) ions.

   a) Describe what the teacher should do to check the solution contains iron(III) ions. (2)

   b) Describe what the teacher should do to check the solid used to make the iron(II) sulfate solution contains iron(II) ions. (3)

   c) Explain, in terms of electrons, the difference between an iron(II) ion and an iron(III) ion. (1)

Data analysis

Example

1 The following table shows the reactions of five metals A–E with water and dilute hydrochloric acid.

Metal	Reaction with water	Reaction with dilute hydrochloric acid
A	no reaction	no reaction
B	bubbles of gas form on the metal surface	produces large volume of gas
C	no reaction	produces bubbles of gas
D	floats and rapidly produces bubbles of gas	produces large volume of gas
E	no reaction	produces bubbles of gas slowly

a) Metals B, C, D and E produce a gas. Describe the chemical test for this gas. (1)

b) Write the letters of the metals in order of their reactivity. Start with the most reactive metal. (4)

c) One of the metals is copper. Which metal is this? (1)

(Total for question = 6 marks)

Student 1 response Total 3/6	Marker comments and tips for success
a) Use limewater. It will turn cloudy white. O	This is the test for carbon dioxide gas. Make sure you know the gases given off by metals and carbonates with acids and the tests for each gas. You should know that the gas here is hydrogen, so the test should be the test for hydrogen.
b) D, B, E, C, A ✓ ✓ ✓	Only three metals are in the right order. Use the reaction with water to get a starting order, and then use the reaction with acid to order the metals that didn't react with water, or have the same description in the water column. Read the descriptions carefully as often only one word is different or extra to show the order.
c) E O	A is copper. Copper is very unreactive, and will only react with concentrated acids.

Student 2 response Total 4/6	Marker comments and tips for success
a) Collect some in a test tube, add a flaming spill and it should burn with a 'pop'. ✓	1 mark for describing the correct test for hydrogen and the positive result.
b) A, E, C, B, D ✓	1 mark only. The order is correct but the metals are written with the *least* reactive first. This means only the middle one is correct. Make sure you complete lists like this as the question asks. The answer loses 3 marks it could have gained.
c) A ✓	Correct. You should remember that unreactive metals such as copper and lead do not react with acids.

2 Chemistry of the elements

Practice questions

2 A student wanted to find out the reactivity of four metals: magnesium, copper, iron and zinc. The student reacted each metal in turn with solutions of the other metal sulfate and observed what happened. Here are the results:

	Metal			
Solution	Magnesium	Copper	Iron	Zinc
magnesium sulfate	X	no reaction	no reaction	no reaction
copper sulfate	blue colour fades	X	blue colour fades	blue colour fades
iron sulfate	green colour fades	no reaction	X	green colour fades
zinc sulfate	silver magnesium goes grey	no reaction	no reaction	X

    a) Explain why some results boxes have an X in them. *(1)*

    b) What is the name given to the type of reaction where one metal removes another metal from its salt? *(1)*

    c) Which metal is most reactive? Explain your answer. *(2)*

    d) Which metal is least reactive? Explain your answer. *(2)*

    e) Write a chemical equation for the reaction of magnesium with iron sulfate ($FeSO_4$) with state symbols. *(2)*

3 Bromine, chlorine, and iodine are known as the halogens. They can all be dissolved in water to make bromine water, chlorine water and iodine water.

A student wanted to investigate which of the three halogens was most reactive. She reacted each colourless halogen water with solutions of sodium chloride, sodium bromide and sodium iodide. She noted any colour changes and recorded her results in a chart.

	Colour of solution after adding halogen water		
Solution	Chlorine water	Bromine water	Iodine water
sodium chloride		colourless	colourless
sodium bromide	orange		colourless
sodium iodide	brown	brown	

    a) Name, if any, the solutions that reacted with bromine water. *(1)*

    b) Name, if any, the solutions that reacted with iodine water. *(1)*

    c) Use information from the table to list bromine, chlorine and iodine in order of their reactivity with each other. Place the most reactive first. *(1)*

    d) Chlorine water reacts with sodium bromide solution. Write a word equation for the reaction. *(1)*

    e) Chlorine water reacts with sodium iodide. Copy and complete the ionic equation for the reaction. *(2)*

        $Cl_2(aq) + \_\_\_ I^-(aq) \rightarrow _____ + _____$

    f) Explain in terms of electrons the reaction between bromine water and sodium iodide. *(2)*

4 Displacement reactions between metals and metal oxides can be useful. The reaction of iron oxide (Fe_2O_3) with aluminium metal can be used to weld metal railway tracks together.

    $\_\_\_ Al(s) + Fe_2O_3(s) \rightarrow _____ + _____$

a) Copy and complete the equation for the reaction. (2)

b) This is an example of a redox reaction.

  i) Name the substance that is oxidised. (1)

  iii) Name the oxidising agent. (1)

c) Here is some information about the displacement reactions between the metals aluminium, copper, sodium and zinc.

Metal	Metal oxide	Does the metal displace the other metal in the metal oxide?
sodium	aluminium oxide	yes
zinc	copper oxide	yes
copper	aluminium oxide	no
aluminium	zinc oxide	yes

  i) List the four metals in order of reactivity. (3)

  ii) What other piece of information would you need to be able to place iron in your order of reactivity? (1)

  iii) Explain your answer to part **ii)** (1)

5 A student wanted to find the order of reactivity of the first three Group 1 and the first three Group 2 metals.

The student reacted all the metals with water. Here are the results.

Group 1 metal	Reaction with water	Group 2 metal	Reaction with water
lithium	floats and produces hydrogen gas	beryllium	no reaction
sodium	melts, floats and produces hydrogen gas	magnesium	produces hydrogen gas slowly
potassium	melts, floats, burns with lilac flame and produces hydrogen gas rapidly	calcium	floats and produces hydrogen gas

a) i) Use the data in the table to identify the trend of reactivity for both groups. (2)

  ii) Write a chemical equation for the reaction of lithium with water. (2)

b) The student could not decide on the order of the metals calcium, lithium, magnesium, and sodium. To help decide the order, the student decided to heat each metal in turn with the oxides of the other three metals.

Here are the results:

Metal	Reaction with metal oxide			
	Sodium oxide	Lithium oxide	Magnesium oxide	Calcium oxide
calcium	no reaction	no reaction	no reaction	
lithium	no reaction		vigorous reaction	vigorous reaction
magnesium	no reaction	no reaction		no reaction
sodium		vigorous reaction	vigorous reaction	vigorous reaction

  i) Explain why the student decided to react the metals with the metal oxides instead of solutions of the metal salts. (1)

  ii) Use data from the two tables to decide on the order of the reactivity of the six metals. List them with the most reactive metal first. (5)

  iii) Write a chemical equation for the reaction of lithium with magnesium oxide. (2)

  iv) Which substance is reduced in the reaction between lithium and magnesium oxide? Explain your answer. (2)

2 Chemistry of the elements

Longer-answer questions

Example

1 Describe how the apparatus shown can be used to make a pure sample of oxygen.

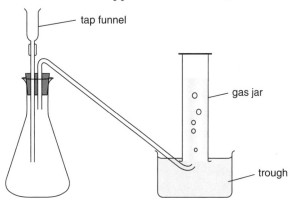

Your answer should refer to:

- the chemicals needed
- why a method is needed to ensure the gas collected is pure oxygen. (6)

(Total for question = 6 marks)

Student 1 response Total 5/6	Marker comments and tips for success
Put three or four spatulas of magnesium(IV) oxide into the flask. Use the tap funnel to add hydrogen peroxide ✓ a drop at a time to the flask. This will make the oxygen gas. Make sure that no air gets into the flask through the tap funnel, as this will make the oxygen impure.	You must take care to use the correct chemical, which is manganese(IV) oxide, particularly when the names are nearly the same.
The oxygen gas will be passing through the water and collecting in the gas jar. The first gas jar will contain air ✓ from the apparatus as well as some oxygen that has been produced so it will not be pure. ✓	The answer gives a clear explanation of where the oxygen gas is collected, and why it is not pure. You should aim for this level of clarity in explanations.
Discard the first two or three gas jars ✓ of air and oxygen before collecting a jar of pure oxygen. ✓	You should give a clear explanation of the reason why an action is done, if it is needed by the question.

Student 2 response Total 2/6	Marker comments and tips for success
Fill the tap funnel with hydrogen peroxide ✓ and add some manganese (IV) oxide ✓ to the flask.	Both chemicals are correctly named, although an indication of quantity would help. Always suggest how much to use.
Open the tap and the oxygen will come out. Collect the pure oxygen in the gas jar.	This answer suggests the oxygen comes out of the tap funnel. Make sure you read your answer and make sure your meaning is clear. The gas comes over to the trough to collect in the gas jar.
You need to make sure the gas is pure otherwise you couldn't breathe it safely.	The student did not read and understand the question. You need to say why the gas will not be pure at the start, but will be pure at the end. Think of possible reasons why the gas might not be pure, and how you can overcome these.

Longer-answer questions

Practice questions

2 Iron is a metal that reacts easily with water and air to rust.

Iron is used in many different ways including making railings for fences.

Iron can be prevented from rusting using grease, paint or galvanising.

Evaluate the use of grease, paint and galvanising for preventing railings for fences from rusting.

You should include an advantage and a disadvantage of each method in your answer. *(7)*

3 Potassium sulfate is a compound that can be made by reacting potassium hydroxide solution with dilute sulfuric acid.

potassium hydroxide + sulfuric acid → potassium sulfate + water

You have been given a sample of pure potassium sulfate crystals.

Describe how you would test the crystals to show they contain potassium sulfate.

In your method you should:

- describe, in detail, the chemical tests you will use to identify the sulfate ion
- describe, in detail, how you will identify the potassium ions in the potassium sulfate
- what the expected results of each test will be. *(6)*

4 Hydrogen chloride is a gas at room temperature

A student dissolved some hydrogen chloride in water. The solution made was acidic.

The student then dissolved the hydrogen chloride in methylbenzene. The solution made was neutral.

Using your knowledge of hydrogen chloride bonding and dissociation explain the student's observations. *(5)*

5 A student wanted to investigate the causes of rusting.

The student set up the following experiment, and left it for one week.

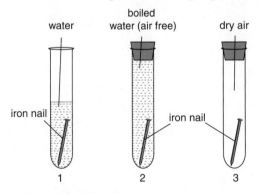

Describe the conditions present in each tube, and explain how this would enable the student to find the causes of rusting. State the results you would expect to see after a week. *(6)*

3 Organic chemistry

Calculations

Example

1. A student investigated a compound similar to ethanol. The student found that this compound contained 59.76% by mass of carbon, 13.33% by mass of hydrogen, and that the rest was oxygen.

   Calculate the empirical formula of this compound. (5)

   (Total for question = 5 marks)

Student 1 response Total 2/5	Marker comments and tips for success
$C, \frac{59.76}{12} = 4.98$ O $H, \frac{13.33}{1} = 13.33$ O So ratio $= \frac{13.28}{4.98} = 2.67$ or $1:2.67$ or $3:8$ ✓ So formula is C_3H_8 ✓	Always read the question carefully. This answer does not include oxygen in the calculation. As a result, the suggested formula is correct for carbon and hydrogen, but has no oxygen. On an *error carried forward* the answer gains 2 marks.

Student 2 response Total 5/5	Marker comments and tips for success
$O = 100 - 59.76 - 13.33 = 26.56\%$ ✓ C is $\frac{59.76}{12} = 4.98$, H is $\frac{13.33}{1} = 13.33$, O is $\frac{26.56}{16} = 1.66$ ✓ Divide by 1.66 ✓ to give $C = 3, H = 8, O = 1$ ✓ Formula is C_3H_8O ✓	1 mark for correctly calculating the percentage of oxygen and 1 mark for dividing by the relative atomic masses. Finding the simple ratio gains 1 mark, with the final mark for the correct empirical formula. You should try to lay out your calculation answers like this one. Allow a line for each step, and clearly state what you are doing.

Practice questions

2. A student investigated a compound similar to ethanol. The student found that this compound contained 37.8% by mass of carbon, 50.4% by mass of oxygen, and that the rest was hydrogen.

   Calculate the empirical formula of this compound. (5)

3. A student investigated a compound containing hydrogen, carbon and oxygen. The compound contained 40.0% by mass of carbon, 6.7% by mass of hydrogen, and 53.3% by mass of oxygen.

   Calculate the empirical formula of this compound. (4)

4. A scientist investigated a hydrocarbon with formula mass 112. The hydrocarbon had 85.7% by mass of carbon and 14.3% by mass of hydrogen.

   a) Calculate the empirical formula of the hydrocarbon. (4)

   b) Calculate the formula of the hydrocarbon. (3)

5. A student investigated a compound. The compound had 56.8% by mass of chlorine, 38.4% by mass of carbon and the rest was hydrogen.

   Calculate the empirical formula of this compound. (5)

Structure, properties and reactions of alkanes and alkenes

Structure, properties and reactions of alkanes and alkenes

Example

1 Methane is the simplest hydrocarbon. The table below shows some information about methane molecules.

	Methane	Ethane
molecular formula	CH_4	
displayed formula	H—C—H with H above and H below (H, |, H—C—H, |, H)	
dot and cross diagram of bonding	H •× C •× H with H above and H below (shown with dots and crosses)	
state at room temperature	gas	

a) Copy and complete the table to show the same information for ethane as is shown for methane. *(4)*

b) What is a hydrocarbon? *(1)*

c) Methane and ethane are part of a homologous series. What is a homologous series? *(1)*

d) Methane burns to produce water vapour. Name the other three possible products of the combustion of methane, and state the conditions necessary for each product to be made. *(3)*

(Total for question = 9 marks)

Student 1 response Total 0/9	Marker comments and tips for success
a) 	Ethane	 molecular formula	C_2H_4 O displayed formula	H—C—H—C—H (with H above and below each C, and O at end) dot and cross diagram of bonding	H •× C—H—C •× H (with H above and below each C, and O at end) state at room temperature	liquid O	The molecular formula is wrong. C_2H_4 is *ethene*, not *ethane*. Remember that changing the 'a' to an 'e' gives a different compound. The displayed formula and the dot and cross diagram are incorrect and do not follow the molecular formula given in the answer. Remember the rule that carbon always makes four bonds, and hydrogen only one. This means that hydrogen only joins to one other atom and cannot form a chain as shown in the two diagrams.
b) A carbohydrate in chemistry. O	This is a common error. Carbohydrates contain carbon, hydrogen and oxygen, and hydrocarbons only contain carbon and hydrogen. You must be sure about the difference.

3 Organic chemistry

Student 1 response — Total 0/9	Marker comments and tips for success
c) A group of compounds that are the same. ○	This starts to explain, but doesn't express clearly that compounds in a homologous series have the same basic formula that increases by a single unit moving along the series. Take time to think about how you will answer before starting to write.
d) Carbon dioxide and carbon monoxide	Insufficient for a mark. Read the question carefully – you are asked for three products and the conditions necessary for each product to be made, so there is 1 mark for each product with the appropriate conditions.

Student 2 response — Total 7/9		Marker comments and tips for success
a)		Full marks. The formula is correct and so are the displayed formula and dot and cross diagram. The answer shows a clear understanding of carbon's need for four bonds, and hydrogen's one bond.
	Ethane	
molecular formula	C_2H_6 ✓	
displayed formula	H H | | H–C–C–H ✓ | | H H	
dot and cross diagram of bonding	H ו•× C ×•ו× C ×•• H ✓ (with H above and below each C)	
state at room temperature	gas ✓	
b) A compound containing only carbon and hydrogen ✓		Correctly explains the elements present in a hydrocarbon.
c) A group of compounds where each compound is different to the next by the same amount. ✓ In this case CH_2.		Gives the correct definition, for 1 mark. The two named examples are used to give the repeat unit of the alkane homologous series.
d) Carbon dioxide when there is lots of oxygen. ✓ Carbon monoxide when there is no oxygen. ○		Only 1 mark out of 3. Read the question carefully to find the number of answers wanted. Carbon dioxide and the conditions are right, but the conditions for carbon monoxide are wrong. There has to be some oxygen present for the methane to burn to make carbon monoxide and carbon.

Practice questions

2 Alkenes are a homologous group with the general formula C_nH_{2n}. Ethene has the molecular formula of C_2H_4.

   a) Draw diagrams to show:

      i)   the displayed formula of ethene   *(1)*

      ii) a dot and cross diagram of an ethene molecule.   *(1)*

   b) C_3H_6 is another member of the alkenes. Draw the displayed formula of C_3H_6. Write the name of the compound beneath it.   *(2)*

   c) Draw the displayed formula of the straight-chain alkene with four carbon atoms. Name the compound, and give its molecular formula.   *(3)*

   d) What would be the molecular formula of the alkene with seven carbon atoms?   *(1)*

   e) Describe how adding bromine water to a sample of hydrocarbon would allow you to decide if the compound was an alkene or an alkane.   *(3)*

Structure, properties and reactions of alkanes and alkenes

3 A student was using a Bunsen burner to heat some water in a beaker. The Bunsen burner was burning methane gas. The student used a yellow or safe flame. After a few minutes there was a black substance ion the bottom of the beaker.

    a) Name the homologous series of which methane is part. (1)

    b) i) Name the black substance that appeared on the bottom of the beaker. (1)

       ii) Explain why the black substance formed. (2)

       iii) Explain how the student could change the apparatus to avoid the black substance forming. (1)

    c) Methane reacts with bromine. Part of the equation for the reaction that occurs is:

    $CH_4 + Br_2 \rightarrow$ _____ + _____

       i) Copy and complete the equation. (2)

       ii) What conditions are needed for the reaction to take place? (1)

       iii) Name one of the products made. (1)

4 Butane (C_4H_{10}) is used in camping gas. It can exist as two different isomers.

    a) Draw the displayed formulae of the two isomers of C_4H_{10}. (2)

    b) What are isomers? (2)

    c) Butene is a member of the alkene homologous series. It has four carbon atoms.

       i) What is meant by a homologous series? (1)

       ii) State **two** ways butene is different from butane. (2)

    d) A student wanted to test samples of butene and butane to identify the butene. Describe what the student should do, and what the expected differences in the results would be. Butane and butene are both gases at room temperature. (2)

5 Ethane (C_2H_6) and ethene (C_2H_4) are both hydrocarbons with two carbon atoms.

    a) i) Draw dot and cross diagrams to show the bonding in ethane and ethene. (2)

       ii) What is the difference between the bonding in ethane and ethene? (1)

       iii) Ethane is a saturated hydrocarbon, and ethene is an unsaturated hydrocarbon. Explain what is meant by the terms **saturated** and **unsaturated**. (2)

b) The graph below shows the boiling points of five members of the alkane series plotted by number of carbon atoms. Alkanes have the general formula of C_nH_{2n+2}

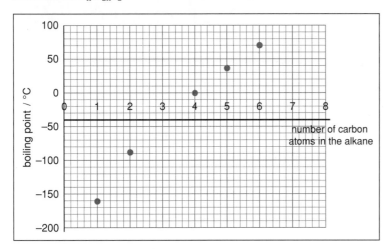

i) What is the formula of the alkane with **six** carbon atoms? *(1)*

ii) What shape of line of best fit would you draw on the graph? Explain your answer. *(2)*

iii) Name and suggest the boiling point of the alkane with **three** carbon atoms. *(2)*

Longer-answer questions

Example

1. Ethanol is very useful as a fuel. It can be made from plant sugars or from ethene obtained from crude oil.

   The displayed formulae of ethanol and ethene are shown.

   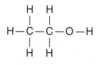
   ethanol

   ethene

   a) Describe how ethanol can be manufactured from ethene. Your answer should include the conditions needed for the reaction. (4)

   b) Ethanol can also be made by the fermentation of plant sugars. Describe the advantages and disadvantages of using ethene instead of plant sugars to make ethanol. (4)

   (Total for question = 8 marks)

Student 1 response Total 2/8	Marker comments and tips for success
a) The <u>ethanol is reacted with steam</u> ✔ with a transition metal catalyst. This produces the ethene. This can be collected over water before being used.	1 mark for saying steam is used. The catalyst is mentioned, but no description of the conditions needed, which are required by the question. When you are asked about conditions for a reaction you need to mention temperature, pressure and name the catalyst where one is used.
b) You cannot grow plants anywhere as they take a long time to grow. ✔ <u>Using plants still produces carbon dioxide and contributes to global warming</u>.	For 4 marks there should be four points. The answer comments on the time taken to grow plants, which is a disadvantage. You should make four points and at least one should be a disadvantage or advantage. All the points are disadvantages. For this type of question you should use as many sentences as there are marks, and state the advantage/disadvantage with an explanation.

Student 2 response Total 5/8	Marker comments and tips for success
a) (steam, 330, atm 60, catalyst?) You would <u>react the ethene with steam</u> ✔ at a <u>temperature of 300°C</u>. ✔ The reaction also needs a catalyst and a <u>high pressure of about 60 times normal atmospheric pressure</u>. ✔ Here is the equation for the reaction: $C_2H_4 + H_2O \rightarrow C_2H_5OH$ The catalyst used is sulfuric acid. O	Writing a quick plan at the start is good way to tackle longer-answer questions. The key points are there, but the name of the catalyst is forgotten. The answer gains 3 marks for correctly describing the reaction. The equation is a help, but the question doesn't ask for one so it gains no mark. At the end the choice of catalyst is wrong, so no mark, but this guess might have been right. You should always try a guess if you know something is needed.
b) Ethene is readily available from crude oil, but you have to wait months for crops to grow. ✔ The reaction is much quicker than having to wait for fermentation by yeast. ✔ You are much more in control.	The question says 'instead of', which means you need to compare the two methods. This answer gets 2 marks only. If you compare the two methods, give an advantage and a disadvantage for each method to get the 4 marks. Don't make the same point as an advantage, then use it again as a disadvantage, for example 'crops take a long time to grow' and 'ethene from oil is available immediately' are the same point.

3 Organic chemistry

Practice questions

2. Plant sugars such as glucose ($C_6H_{12}O_6$) can be converted to ethanol by fermentation.

   Describe how you could make a pure sample of ethanol in the laboratory from glucose. Write a chemical equation for the reaction. You may use a diagram to help you. (6)

3. Ethanol is a chemical that can be used to produce fuel for motor vehicles. There are two methods for producing ethanol. It can be obtained by the reaction of ethene from crude oil. Alternatively it can be produced by fermentation from food crops such as sugar cane.

   Evaluate the factors that should be considered when selecting the method of ethanol production. Suggest with reasons your preferred production method. (6)

4. Crude oil is a source of ethene. Ethene is used to make many different polymers (plastics). Crude oil is becoming increasingly scarce. Scientists have developed a method for making ethene from ethanol so that as crude oil prices rise there will still be an inexpensive source of ethene.

   a) Describe how ethanol is converted to ethene. Your answer should include the conditions needed for the reaction, and a chemical equation. (3)

   b) Give **three** disadvantages of using ethene from ethanol instead of ethene from crude oil to make polymers. (3)

   c) Ethene can be converted to ethanol using a phosphoric acid catalyst. What is a catalyst? (1)

5. Ethanol can be made from plant sugars or ethene from crude oil. One scientist said, 'Whenever possible ethanol should be made from plant material.' A different scientist said, 'Using plant materials to make ethanol to burn in cars will cause the poor to starve, and the rich to travel, we should use crude oil.'

   Use your knowledge of the two methods of obtaining ethanol to evaluate the two statements. State which method of production you would choose with reasons. (6)

4 Physical chemistry

Charges, chemical formulae and equations

Example

1. Copper was first extracted by reduction from copper ores containing copper carbonate in the Middle East. The copper carbonate was heated in fires with wood charcoal to produce impure copper.

   a) Copy and complete the chemical equation. (2)

   ___ $CuCO_3(s) + C(s) \rightarrow$ _____ + _____

   b) Calculate the mass in grams of 1 mole of copper carbonate. (2)

   c) Use the equation to:

   i) give the state of the carbon at the start (1)

   ii) state the number, in moles, of carbon dioxide molecules released when 1 mole of copper is made (1)

   iii) calculate the mass of carbon dioxide released by the reaction of 1 mole of copper carbonate with carbon. (3)

   (Total for question = 9 marks)

Student 1 response — Total 8/9	Marker comments and tips for success
a) $2CuCO_3(s) + C(s) \rightarrow 2Cu(s) + 3CO_2(g)$ ✔✔	The formulae for the products are correct and the balancing is correct.
b) $CuCO_3 = 63.5 + 12 + (3 \times 16)$ ✔ $= 123.5$ ✔	The formula is given in the question. 1 mark for correct substitution of relative atomic masses and 1 mark for correct addition.
c) i) solid ✔	The state symbol provided in the question gives the answer.
ii) 3 moles O	This answer comes from the equation in part a), but 3 moles of carbon dioxide are produced for every 2 moles of copper, not 1 mole of copper. Half the copper means half the carbon dioxide, so divide the 3 moles by 2 to get 1.5 moles.
iii) CO_2 is 44, ✔ so 3×44 ✔ $= 132\,g$ ✔	1 mark for correctly stating the relative formula mass of carbon dioxide as 44. The number of moles in part c) ii) is wrong, but the error is carried forward. The working gains 1 mark for showing the 3 moles multiplied by the formula mass of CO_2. The final mark is gained for correct calculation. If you do not show your working, you will miss out on these marks.

Student 2 response — Total 5/9	Marker comments and tips for success
a) $CuCO_3(s) + C(s) \rightarrow Cu(s) + CO_2(g)$ ✔	1 mark for the correct formulae, but the equation is not balanced. To gain the second mark you must balance the equation. Correct the carbon first to give $2CO_2$. The oxygen now has 3 atoms as reactants, but 4 atoms as products. Increase the oxygen in the reactants by making it $2CuCO_3$. There are now 2 Cu atoms so the products need to be 2Cu. The 6 oxygen atoms provide two more than needed, but there is an extra carbon atom. This is a CO_2 molecule, so you need to change $2CO_2$ to $3CO_2$ and the equation balances.
b) $63.5 + 12 + 16 + 16 = 107.5$ O	The calculation only includes **two** oxygen atoms, not **three**. You must make sure when calculating formula mass that you carefully read and use the subscript numbers.
c) i) solid ✔	The state symbol tells you the state of the chemical.
ii) 1 mole O	This is a guess, as no attempt was made to balance the equation in part a).
iii) $12 + 16 + 16$ ✔ $= 44$ 44×1 ✔ $= 44\,grams$ ✔	1 mark for calculating correctly the formula mass of carbon dioxide. 2 marks on error carried forward for method and calculation.

4 Physical chemistry

Practice questions

2 Ammonium sulfate is an important chemical fertiliser. It is made by reacting ammonia gas with sulfuric acid.

The equation for the reaction is:

$2NH_3(aq) + H_2SO_4(aq) \rightarrow (NH_4)_2SO_4(aq)$

   a) What information in the equation shows that the ammonia gas is dissolved in water before reacting with the sulfuric acid? *(1)*

   b) What amount, in moles, of ammonia reacts with 1 mole of sulfuric acid? *(1)*

   c) One mole of gas has a volume of 24 000 cm³ at room temperature. Calculate the volume of ammonia that needs to be dissolved in water to make 1 mole of ammonium sulfate. *(2)*

   d) Calculate:

   i) the relative formula mass of ammonia *(2)*

   ii) the relative formula mass of ammonium sulfate. *(2)*

   e) Describe how a pure dry sample of ammonium sulfate can be obtained from the $(NH_4)_2SO_4(aq)$. *(2)*

3 A student wanted to make a sample of barium sulfate. The student reacted some barium nitrate solution with dilute sulfuric acid. The equation for the reaction is:

$Ba(NO_3)_2(aq) + H_2SO_4(aq) \rightarrow BaSO_4(s) + 2HNO_3(aq)$

Use information in the equation to help you answer the questions.

   a) Explain how the student could obtain a sample of barium sulfate. *(2)*

   b) Calculate the formula mass of barium sulfate. *(2)*

   c) The student used 0.25 moles of barium nitrate. Calculate the mass, in grams, of barium sulfate that would be made. *(2)*

   d) The student dried the barium sulfate made and then weighed it. The mass was 36.12 g. Use your answer to part c) to help you calculate the percentage yield of the student's experiment. *(2)*

   e) Barium sulfate is used in hospitals to help investigate digestive system problems. Patients are given a barium 'meal' to drink before having an x-ray. The barium 'meal' is a suspension of barium sulfate in water. Barium ions are highly toxic. Suggest why it is safe to drink a barium 'meal'. *(2)*

4 People who go climbing mountains can find it hard to heat up food to eat. Several specialist food companies provide self-heating packs of food. The diagram shows how one type of self-heating food pack works.

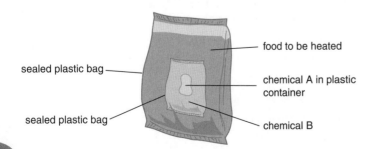

When the climber wants to heat the food the plastic container containing chemical A is broken. This allows chemical A to react with chemical B, releasing a lot of heat energy.

a) What is the name given to reactions that release heat energy? (1)

b) Two chemicals that can be used for the reaction are calcium oxide and water.

Copy and complete the equation for the reaction of calcium oxide with water. (1)

$CaO(s) + H_2O(l) \rightarrow$ _____

c) Calculate the formula mass of calcium oxide. (2)

d) A student investigated the reaction of 3.5 g of calcium oxide in 25 cm³ of water. There was a temperature rise of 35 °C.

   i) Calculate the amount, in moles, of calcium oxide the student used. (2)

   ii) How many moles of water would the calcium oxide react with? (1)

e) Theoretically the reaction should produce the same number of moles of calcium hydroxide as calcium oxide that is used. In the student's reaction only 0.0455 moles of calcium hydroxide were made.

Calculate the percentage yield of the reaction. (2)

5 A student used the reaction of sodium thiosulfate with hydrochloric acid to measure the rate of a chemical reaction.

The equation for the reaction is:

$Na_2S_2O_3(aq) + 2HCl(aq) \rightarrow S(s) + 2NaCl(aq) + SO_2(g) + H_2O(l)$

a) As the reaction proceeds the solution becomes cloudy. Name the substance that turns the solution cloudy. (1)

b) The student used 1.58 g of sodium thiosulfate in the reaction.

   i) Calculate the formula mass of sodium thiosulfate. (2)

   ii) Calculate the amount, in moles, of sodium thiosulfate used. (2)

   iii) State the amount, in moles, of sulfur dioxide produced. (1)

c) One mole of gas has a volume of 24 000 cm³ at room temperature. Calculate the volume, in cm³, of sulfur dioxide that will be produced using 1.58 g of sodium thiosulfate. (2)

6 The presence of water in a substance can be shown using white anhydrous copper sulfate. The equation for the reaction is:

$CuSO_4(s) + 5H_2O(l) \rightleftharpoons CuSO_4 \cdot 5H_2O(s)$

a) What colour is $CuSO_4 \cdot 5H_2O(s)$? (1)

b) How many moles of water will react with 1 mole of anhydrous copper sulfate? (1)

c) Calculate the mass, in grams, of 1 mol of $CuSO_4 \cdot 5H_2O(s)$. (2)

The reaction of anhydrous copper sulfate with water is reversible.

d) What is meant by a **reversible reaction**? (1)

e) What piece of information in the equation tells you the reaction is reversible? (1)

4 Physical chemistry

Practical work 1

Example

1 A student wanted to test some household products to see if they were acidic, alkaline or neutral. She had no universal indicator, but knew she could make an indicator as shown in the diagrams below.

a) The student heated the red cabbage and water for 10 minutes, and allowed the solution to cool.

Describe what she must do next to obtain the indicator solution. (1)

b) Describe how the student could find out the colour of her red cabbage indicator in acids, alkalis and neutral solutions. (3)

c) Copy and complete the table showing the pH scale. (4)

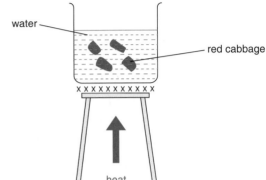

Colour of indicator				Blue	
pH	0	5	7	9	14
Strength of acid / alkali				weak alkali	

d) Describe how you could use universal indicator paper to test the pH of some liquids. (2)

(Total for question = 10 marks)

Student 1 response Total 6/10	Marker comments and tips for success
a) Filter the solution. ✔	1 mark for correctly identifying how to separate the solid red cabbage from the solution.
b) She should get some acids and alkalis ✔ and react them together with the indicator. ✔	The answer is unstructured. Answer this type of question in the form of instructions: 'Do this ...' then 'do that ...', followed by how you would use the results.

c)
 	
  | Colour of indicator | red | orange | green ✔ | blue | blue |
  | pH | 0 | 5 | 7 | 9 | 14 |
  | Strength of acid / alkali | acid | acid | neutral | weak alkali | alkali |
	The answer is too imprecise. The table clearly states the *strength* of the alkali at pH 9. You need to say if the acid or alkali is strong or weak. The answer only describes the pH as acid or alkali, so only the pH 7 column gains the mark for the correct colour and correct description.

d) Put the indicator in each liquid. ✔ Take it out then look at the chart. ✔	To make the instructions really clear, imagine you are telling an 11-year-old student how to use indicator paper. There are 2 marks: dip the paper in the solution, then compare with a pH chart to find the nearest match.

Practical work 1

Student 2 response Total 10/10	Marker comments and tips for success
a) Pass the solution through some filter paper ✔ to remove the red cabbage. Then use the solution.	A good clear answer.
b) Put an equal volume of an acid such as vinegar in one test tube, some oven cleaner which is alkali into another one, and some tap water which should be neutral in a third test tube. ✔ Add enough indicator solution to colour each liquid. ✔ Note the colour of the red cabbage in each tube. ✔ This tells you the colour of the indicator in acid, alkali and neutral solutions.	Full marks. Notice how the answer uses clear sentences and explains each step in order.
c) 	Colour of indicator	red ✔	orange ✔	green ✔	blue	purple ✔	 	pH	0	5	7	9	14	 	Strength of acid / alkali	strong acid	weak acid	neutral	weak alkali	strong alkali		The answer correctly shows the right colours and the strength of the acid and alkali. Make sure you know the colours of these five pH values for universal indicator.
d) Dip a piece of fresh indicator paper in each liquid. ✔ Compare the colour that has developed to a pH chart to find the best pH value. ✔	Full marks for a good answer giving the key steps in order.

Practice questions

2 The gas ammonia dissolves in water to form a solution. This solution turns red litmus paper blue.

  a) i) Explain what the litmus paper test shows about the solution of ammonia in water. *(1)*

  ii) The solution of ammonia in water is sometimes called ammonium hydroxide.

  Write the formulae of the two ions present in ammonium hydroxide. *(2)*

  iii) Which ion has caused the red litmus paper to turn blue? *(1)*

  b) Hydrogen chloride is a gas that dissolves in water. The solution causes methyl orange solution to turn red.

  i) Explain what the methyl orange test shows about the solution of hydrogen chloride in water. *(1)*

  ii) What is a solution of hydrogen chloride in water called? *(1)*

  iii) Write the formulae of the two ions present in the solution of hydrogen chloride in water. *(2)*

  iv) Which ion has caused the methyl orange solution to turn red? *(1)*

  c) A student mixed fresh samples of the solutions of hydrogen chloride and ammonia together. The solutions neutralised each other.

  i) Describe a method the student could use to test that the final solution was neutral. *(2)*

  ii) What is meant by the term **neutral**? *(1)*

  iii) Name the two compounds made by the neutralisation of ammonia solution with hydrogen chloride solution. *(2)*

4 Physical chemistry

3 a) Copy and complete the table below showing the colours of three indicators in acid, alkaline and neutral solutions. (3)

Indicator	Colour of indicator in		
	Acidic solutions	Neutral solutions	Alkaline solutions
methyl orange			
phenolphthalein			
litmus			

b) Explain which indicator to use to show that a solution has just become alkaline. (2)

c) Explain which indicator to use to show that a solution has changed from a strong acid to neutral. (2)

d) A student added enough sodium hydroxide solution to dilute hydrochloric acid to neutralise the solution.

The equation for the reaction is:

$NaOH(aq) + HCl(aq) \rightarrow NaCl(s) + H_2O(l)$

i) Explain how the student could check the solution was neutral using some universal indicator paper. (2)

ii) What is meant by a **neutral solution**? (1)

iii) The student's friend said that the equation could be simplified to show just the ions taking part in neutralisation. Copy and complete the simplified ionic equation for neutralisation. (1)

_____ + _____ $\rightarrow H_2O(l)$

4 A student placed three different dilute acids in three test tubes. Each tube had five drops of universal indicator solution added to it. He then added magnesium oxide to the sulfuric acid, calcium carbonate to the nitric acid, and magnesium to the hydrochloric acid, as shown in the diagram. The student added enough magnesium oxide, calcium carbonate and magnesium to change the colour of the universal indicator.

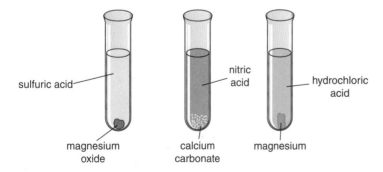

a) Describe what the student would see in:

i) the test tube with magnesium oxide and sulfuric acid (2)

ii) the test tube with calcium carbonate and nitric acid. (3)

b) One of the gases produced was hydrogen. Explain how you would test to show that a gas was hydrogen gas. (1)

c) Name all the products of:
   i) the reaction between magnesium oxide and sulfuric acid (2)
   ii) the reaction between calcium carbonate and nitric acid (3)
   iii) the reaction between magnesium and hydrochloric acid. (2)

d) When the reactions had finished, the student then tested the pH of each solution using a pH meter. Here are his results.

pH of the solution after the reaction of		
Magnesium oxide and sulfuric acid	Calcium carbonate and nitric acid	Magnesium and hydrochloric acid
8.3	7.05	1.3

   i) The student measured one pH with greater precision than the others. State, with reasons, which one this was. (2)
   ii) State, with reasons, which acid was added to **excess** in the experiment. (2)

5 A student wanted to prepare some carbon dioxide in the laboratory. She set up the apparatus shown.

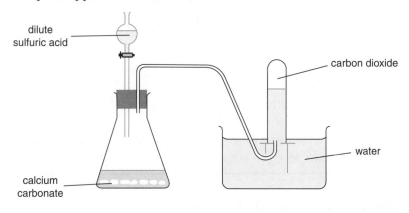

After a few bubbles of gas the reaction stopped. Her friend said this was because she had used the wrong acid.

a) Which acid should the student have used to make the carbon dioxide gas? (1)

b) Explain why the reaction with sulfuric acid stopped after a few bubbles were produced. (3)

c) How could the student test the gas produced to show it was carbon dioxide gas? (2)

d) The student wanted to know how quickly the carbon dioxide gas was being produced. At first she counted bubbles, but she soon lost count.
   i) Suggest a change to the apparatus that would enable the student to accurately record the volume of gas being produced in one minute. (1)
   ii) What other piece of equipment would the student need to find out how much gas was produced in one minute? (1)

4 Physical chemistry

Practical work 2: making salts

Example

1. A student wanted to make some copper sulfate. Copper sulfate is a soluble salt. The student had some copper oxide and reacted it with dilute sulfuric acid.

   a) Describe the steps needed for the student to obtain a pure neutral sample of copper sulfate crystals. (4)

   b) Describe how the student could obtain some dry copper sulfate crystals from the copper sulfate solution. (2)

   c) Describe a chemical test the student could do to show that the crystals obtained contained sulfate ions. (3)

   (Total for question = 9 marks)

Student 1 response — Total 7/9	Marker comments and tips for success
a) Heat 25 cm^3 of dilute sulfuric acid in a beaker over a Bunsen burner. Add a spatula at a time of copper oxide. ✔ Stir between each addition and keep adding until no more will dissolve ✔. The solution will look blackish, but the liquid will be blue. Filter the hot liquid, ✔ and place the filtrate in an evaporating dish. Heat until one-third is left, then allow to cool. ✔	This is the standard of answer you should aim for. You should remember there are three methods for making a salt. You either use precipitation for insoluble salts, and wash and filter the precipitate, titration for alkalis or react a solid with an acid. Metals, metal carbonates and metal oxides all react with the acid.
b) Filter to obtain the crystals. ✔	Filtering is correct but the question asked for dry crystals. Make sure you read the question carefully.
c) Add the crystals to some dilute hydrochloric acid and barium chloride solution. ✔ It should give a white precipitate. ✔	You need to remember all the steps in a test. For this test the crystal should first be dissolved, so 1 mark is lost.

Student 2 response — Total 4/9	Marker comments and tips for success
a) Add some of the acid to the copper oxide. When it has all dissolved filter the liquid. ✔ <u>Boil the liquid you get until it is dry</u>. You should have some copper sulfate crystals. ✔	The answer explains how to obtain the crystals but not how to make sure all the sulfuric acid has reacted. The solution is almost certainly still acidic. You should always make sure the acid has completely reacted by checking the pH or using an excess of the other chemical. Never boil the solution until it is dry.
b) Take the crystals out of your basin. They should be dry. If not dab them with a paper towel. ✔	When answering questions on practicals imagine you are doing the practical. What would you do next? How would you do it? The solution should be filtered before drying the crystals.
c) Add silver nitrate to get a white precipitate. O	This is the test for halide ions. Make sure you know all the chemical tests in the specification.

Practical work 2: making salts

Practice questions

2 Some salts can be made by a precipitation reaction. Calcium sulfate (CaSO$_4$) is one of them.

   If sodium sulfate solution is added to calcium nitrate a precipitate of calcium sulfate is made.

   The equation for the reaction is:

   Na$_2$CO$_3$(aq) + Ca(NO$_3$)$_2$(aq) → CaSO$_4$(s) + 2NaNO$_3$(aq)

   a) What is a precipitate? (1)

   b) Describe how a pure, dry sample of calcium sulfate can be made using this reaction. (4)

   c) 0.01 mol of sodium sulfate is reacted with excess calcium nitrate.

      i) What is meant by the term **excess**? (1)

      ii) Calculate the amount, in moles, of calcium sulfate that should be made. (1)

      iii) Calculate the amount, in moles, of sodium nitrate that should be made. (1)

      iv) Calculate the mass, in grams, of calcium sulfate that should be made. (3)

   d) A student carried out the reaction using 0.01 mol of sodium sulfate. She weighed the dry calcium sulfate and found that it weighed 0.95 g.

   Use your answer to part **c) iv)** to calculate the percentage yield of the student's experiment. (2)

3 A student wanted to make some zinc chloride (ZnCl$_2$). The student had some zinc oxide (ZnO).

   a) Name the acid that the student should react with the zinc oxide to make zinc chloride. (1)

   b) Describe the steps needed for the student to obtain a pure dry sample of zinc chloride. (5)

   c) Write a chemical equation for the reaction. (2)

4 A student wanted to make some sodium sulfate (Na$_2$SO$_4$). Here is the method the student used.

   The student added an excess of sodium hydroxide (NaOH) to an acid.

   The solution was evaporated to one-third the original volume.

   The sodium sulfate crystals were obtained by filtering.

   a) Name the acid that the student used. (1)

   b) The student discovered that the crystals were alkaline instead of neutral. Explain why this had happened. (3)

   c) Suggest **two** changes to the experimental method that the student should do to make sure that the only sodium sulfate was in the crystals. (2)

   d) Write a chemical equation for the reaction. (2)

4 Physical chemistry

5 The table shows three different methods for making salts.

Method 1	Method 2	Method 3
add an alkali to an acid	add an **excess** of carbonate, metal or metal oxide to an acid	mix solutions of two salts together
check the solution is neutral	filter the solution	filter, and wash the solid on the filter paper
evaporate solution to obtain salt	evaporate solution to obtain salt	dry solid to obtain salt

a) i) Explain, with a reason, which method would be best to make a sample of copper sulfate. (2)

ii) Suggest the **two** chemicals you would use to make copper sulfate. (2)

b) i) Explain, with a reason, which method would be best to make a sample of potassium chloride from an alkali. (2)

ii) Suggest the **two** chemicals you could use to make the sample of potassium chloride. (2)

c) i) Explain, with a reason, which method would be best to make a sample of lead iodide. (2)

ii) Suggest the two chemicals you could use to make lead iodide. (2)

d) Explain why nitrates can only be made using method 1 or method 2. (2)

6 A student wanted to make some nickel chloride crystals from some nickel oxide. Here is the student's proposed method.

Step 1 – Warm 50 cm^3 of dilute sulfuric acid in a 250 cm^3 beaker.

Step 2 – Add three spatulas of nickel oxide.

Step 3 – When all the nickel oxide has dissolved filter the solution.

Step 4 – Place the solution in an oven until it is dry.

a) The student has several problems with the method. Describe **two** problems, and what the student should do to correct these problems. (4)

b) Describe a chemical test the student could do to show that the crystals obtained contained chloride ions. (3)

Data analysis

Example

1 Hydrogen peroxide (H_2O_2) decomposes to water and oxygen. A student investigated adding manganese(IV) oxide to hydrogen peroxide by measuring the volume of oxygen gas given off over a period of ten hours by 50 cm³ of hydrogen peroxide. Here are the results:

Time (hours)	Volume of oxygen gas / cm³		
	No manganese(IV) oxide	With 1 g of manganese(IV) oxide	With 2 g of manganese(IV) oxide
0	0	0	0
2	5	495	505
4	10	785	775
6	15	945	935
8	20	985	975
10	25	990	985

a) What do we call a substance that alters the rate of a chemical reaction and is unchanged at the end? *(1)*

b) Use data from the table to describe the difference, if any, between the reaction with 1 g of manganese(IV) oxide and without manganese(IV) oxide. *(2)*

c) Use data from the table to describe the difference, if any, between the reaction with 1 g of manganese(IV) oxide and with 2 g of manganese(IV) oxide. *(2)*

d) What conclusion should the student make about using manganese(IV) oxide on the effect of the rate of the reaction from the table? *(2)*

(Total for question = 7 marks)

Student 1 response — Total 6/7	Marker comments and tips for success
a) catalyst ✔	Correct. Make sure you know the correct scientific terms.
b) After every time there is far more oxygen ✔ made with the manganese(IV) oxide than without.	1 mark for stating that more oxygen is made with manganese(IV) oxide. To gain the second mark the answer needs to refer to data from the 1 g column of the table.
c) Although all the figures are different they are in the <u>same sort of area</u> ✔, and they both seem to slow up towards the end. I think there is <u>no real difference</u> between the effect. ✔	Poorly expressed but right. Remember that experimental data will have random errors. You should concentrate on the trends rather than the exact figures. Saying 'no real difference' is quite correct from the data.
d) It increases the rate of reaction. ✔ It probably doesn't matter how much you use. ✔	A good conclusion made by comparing the data about the effect of the catalyst, and the mass of catalyst used.

4 Physical chemistry

Student 2 response	Total 1/7	Marker comments and tips for success
a) manganese(IV) oxide ○		This is the name of the catalyst. Make sure you answer the question asked.
b) It goes faster. ✔		1 mark only; the answer does not use the data in any way to support the difference described. You should always refer to data in some way, such as quoting the volumes after 4 hours as a comparison.
c) It starts off better with 2 g but by the end its better at 1 g. ○		The answer compares individual figures without looking at the trend. You should realise that experimental data always shows some random error or variation so the figures will be different, but nearly the same. The trend is the same for both columns. Do not focus on individual results – use the trend.
d) The more you use the better. ○		This conclusion shows the student has not looked carefully at the data. Look at all the data, not just the first row. Over time, doubling the mass of the catalyst makes no difference to its effectiveness.

Practice questions

2 Ammonia gas (NH_3) can be made by reacting nitrogen gas from the air with hydrogen gas from natural gas.

The equation for the reaction is:

$N_2(g) + 3H_2(g) \rightleftharpoons 2NH_3(g)$

The reaction is exothermic

a) The reaction is a dynamic equilibrium. What is a dynamic equilibrium? *(2)*

The table shows the effect of increasing temperature on the percentage of ammonia that is made.

Temperature / °C	200	300	400	500
percentage of ammonia made after 1 day without a catalyst at normal pressure	15	7	0	0
percentage of ammonia made after 7 days without a catalyst at normal pressure	51	18	6	2
percentage of ammonia made after 1 day with a catalyst at normal pressure	50	19	5	3
percentage of ammonia made after 1 day with a catalyst at 50 atm pressure	62	23	9	3

b) What is the effect on the equilibrium of increasing the temperature at a constant pressure? *(1)*

c) The reaction uses a catalyst. What is the purpose of the catalyst? *(1)*

d) What evidence is there that the catalyst has been effective? *(2)*

e) Explain what will happen to the number of molecules of ammonia gas if:

  i) the pressure is increased at a constant temperature *(1)*

  ii) the temperature is increased at a constant pressure *(1)*

  iii) some of the molecules of ammonia are removed. *(1)*

3 Sulfur dioxide reacts with oxygen to produce sulfur trioxide. The equation for the reaction is:

$2SO_2(g) + O_2(g) \rightleftharpoons 2SO_3(g)$

The table shows the conversion percentages at different temperatures.

Temperature / °C	350	400	450	500
% converted to SO_3 / g	98	97	96	92

a) What trend is shown about the temperature and conversion to sulfur trioxide of sulfur dioxide? *(1)*

b) Explain, using data from the table, if the reaction is exothermic or endothermic. *(2)*

c) At 100 °C the reaction very slowly converts all the sulfur dioxide to sulfur trioxide. In industry the temperature chosen is 450 °C despite the lower yield. Suggest why. *(1)*

d) A student suggested that increasing the pressure might improve the yield. Explain why this would be successful. *(2)*

e) Even at 450 °C the reaction is very slow. Explain what could be done to speed up the rate of the reaction at 450 °C whilst maintaining the yield. *(2)*

4 The yield of ammonia gas from the reaction of nitrogen and hydrogen can be affected by the pressure at which the reaction takes place.

The equation for the reaction is:

$N_2(g) + 3H_2(g) \rightleftharpoons 2NH_3(g)$ the reaction is exothermic

Pressure in atm	100	200	300	400
% converted to NH_3 /g at 500 °C	10	20	24	27

a) What does the $\rightleftharpoons$ symbol in the equation show about the reaction of nitrogen and hydrogen to make ammonia? *(1)*

b) Describe how changing the pressure affects the yield of ammonia. Explain why. *(2)*

c) How will reducing the temperature affect the yield of ammonia at 100 atm? Explain why. *(2)*

d) The reaction is very slow. Explain the benefit of using a catalyst as well as a high temperature. *(2)*

4 Physical chemistry

■ Working with graphs

Example

1 A student wanted to find out how changing the temperature, concentration and particle size affects the rate of a chemical reaction.

The student decided to react excess sodium carbonate with hydrochloric acid. He measured the volume of carbon dioxide gas produced against time. He did one reaction, and used this to compare with his other reactions. Graph B shows his original experiment. Graphs A, C and D show his other experiments.

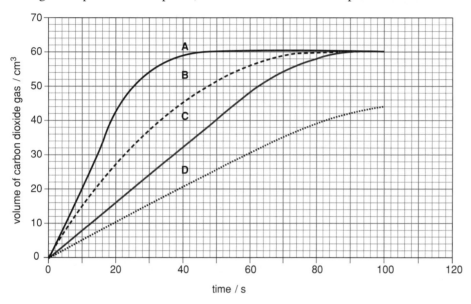

a) Explain why lines A, B and C all rise to a maximum but then do not change. (2)

b) Explain which graph has the fastest rate of reaction. (2)

c) Explain which graph could show the reaction taking place at a lower temperature. (2)

d) Explain which graph shows the reaction has not been completed. (2)

(Total for question = 8 marks)

Student 1 response Total 7/8	Marker comments and tips for success
a) At the <u>maximum point all the sodium carbonate has reacted.</u> ✔	1 mark only. Make sure you know which substance is in excess. Here the substance in excess is sodium carbonate, so it is the hydrochloric acid that has run out.
b) A ✔ as it has the steepest curve. ✔	Correct. Remember that in rate graphs, the steepest curve has the fastest reaction.
c) C ✔ as the reaction is slower. ✔	Correct. You need to know how temperature affects the rate of a reaction. Curve C is less steep and therefore slower than B, so it is the most likely lower temperature reaction. As D does not reach the same height as A, B, or C, it is likely to be at a different concentration rather than temperature.
d) D ✔ as the reaction line has not flattened out. ✔	Correct. The flat part of the graph shows that the reaction has stopped. If it does not become flat, then the reaction is still happening.

Working with graphs

Student 2 response Total 2/8	Marker comments and tips for success
a) The reaction has finished. ✔	1 mark only. You need to show you know that a maximum is reached when at least one of the reactants has been completely used up. You must also name the reactant to gain the second mark.
b) C, it has the lowest curve that reaches the top. O	To answer this type of question you need to understand that the steeper the curve, the faster the reaction. Carefully look at the curves and decide which is fastest, slowest, and other features before attempting the question.
c) A O	With this 'explain' question you need to give both a choice of curve and a reason for your choice to gain both marks. If you can't explain your choice, it may be the wrong one.
d) D ✔ but it may just have finished.	1 mark only. The answer should state that the curve has not levelled off. Concentrate on what you can see on the graph, not what might be about to happen.

Practice questions

2 A student did an experiment to find the effect of increasing the temperature on the rate of a reaction. She reacted 0.5 g magnesium ribbon with 50 cm³ of dilute hydrochloric acid and recorded the time taken to collect 100 cm³ of hydrogen gas.

The student repeated the experiment at five different temperatures.

The graph shows the results she obtained.

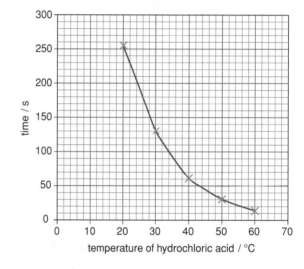

a) Why did she use five different temperatures? *(1)*

b) What conclusion should be made from the graph about the rate of reaction and temperature? *(1)*

c) Explain, in terms of collisions, why changing the temperature affects the rate of a reaction. *(3)*

d) State the time, in seconds, that would be needed to produce 100 cm³ of hydrogen gas at 35 °C. *(1)*

e) The student wanted to double the surface area of the magnesium ribbon and repeat the experiment at 40 °C.

   i) Suggest how long it would take to collect 100 cm³ of hydrogen gas if the magnesium's surface area was doubled and there was an excess of hydrochloric acid. *(1)*

   ii) Explain, in terms of collisions, your answer to part **i)**. *(2)*

3 Hydrogen gas is made when zinc metal is reacted with dilute sulfuric acid.

The equation for the reaction is:

$Zn(s) + H_2SO_4(aq) \rightarrow ZnSO_4(aq) + H_2(g)$

Graph A shows the volume of gas produced over 100 seconds when the reaction takes place. Graph B shows the same reaction but with the addition of a spatula of copper sulfate to the zinc. The copper sulfate is unchanged at the end of the reaction.

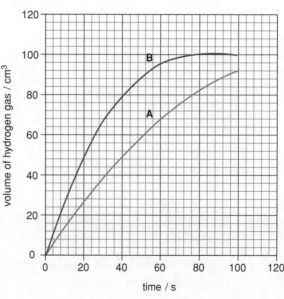

59

a) Explain why the copper sulfate has been added to the mixture. (2)

b) Describe the effect of adding the copper sulfate to the reaction. (2)

The diagram shows one possible explanation of how a catalyst works.

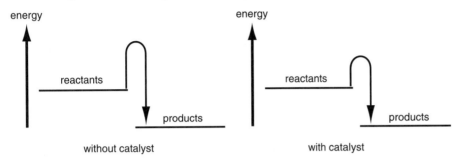

c) Copy the first of the two diagrams and mark on the activation energy. (1)

d) What is meant by **activation energy**? (2)

e) Explain, in terms of activation energy, how a catalyst works. (1)

f) Explain, in terms of collisions, how altering the activation energy can increase the rate of a chemical reaction. (3)

4 A student investigated the effect of changing the particle size on a chemical reaction. He reacted an excess of marble chips ($CaCO_3$) with dilute hydrochloric acid. The reaction produced carbon dioxide gas. The student followed the rate of the reaction by measuring the loss of mass as shown in the diagram.

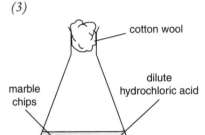

a) What is meant by the term **an excess**? (1)

The student used two samples of marble chips with different chip sizes.

The graph shows the results he obtained.

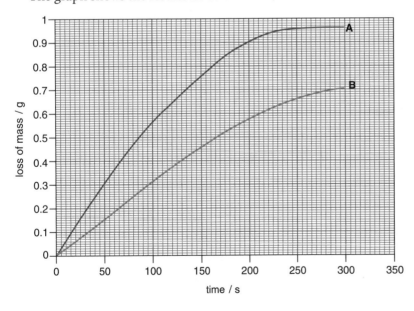

b) i) Explain which graph shows the marble chips with the larger surface area. (2)

ii) Explain, in terms of collisions why altering the surface area of a reactant changes the rate of reaction. (3)

c) Use the graph to find the loss of mass, in grams, in experiment B after 120 s. *(1)*

d) i) Which reaction went to completion? *(1)*

   ii) Explain why you chose this reaction. *(2)*

   iii) State after what time, in seconds, the reaction was complete. *(1)*

5 A student investigated how concentration affected the rate of a chemical reaction. The student used sodium thiosulfate and dilute hydrochloric acid in a disappearing cross experiment. The reaction produces a yellow precipitate of sulfur that makes a cross placed under the reaction flask disappear. The student timed how long it took for the cross to disappear when he looked down on the flask.

The student used three different concentrations of dilute hydrochloric acid solution in the experiment.

At the end of the lesson the student's teacher said it would have been better to use five different concentrations.

a) Explain why it would have been better to use five different concentrations of hydrochloric acid. *(2)*

The graph shows the results the student obtained.

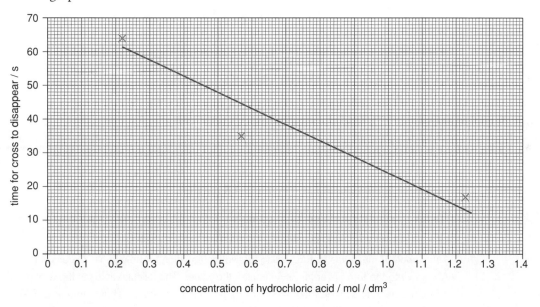

b) Which three concentrations of hydrochloric acid did the student use? *(1)*

c) Explain why the line does not pass through any of the points. *(2)*

d) From the graph, calculate the time it would take for the cross to disappear if the student used 1 mol/dm³ hydrochloric acid. *(1)*

e) What is the relationship between concentration and the time for the cross to disappear? *(1)*

f) Explain, in terms of collisions, your answer to part **e**). *(3)*

4 Physical chemistry

■ Calculations 1

Example

1 A student mixed 25 cm³ of 1 mol/dm³ hydrochloric acid and 25 cm³ of 1 mol/dm³ sodium hydroxide solution together in an insulated beaker. The reaction is exothermic.

The equation for the reaction is:

$HCl(aq) + NaOH(aq) \rightarrow NaCl(s) + H_2O(l)$

The student recorded the temperatures of the hydrochloric acid and sodium hydroxide solution before the experiment and the highest temperature reached during the reaction.

The table shows the results obtained.

Temperature of the hydrochloric acid / °C	Temperature of the sodium hydroxide solution / °C	Highest temperature of the final solution / °C
21	21	28

a) Calculate, in °C, the temperature rise for the reaction. (1)

b) Calculate, in joules, the heat energy released during the reaction. Assume the specific heat capacity of the solution is 4.2 J/g/°C, and the mass of 1 cm³ of solution is 1 g. (3)

c) Calculate the amount, in moles, of hydrochloric acid the student used. (2)

d) Calculate, in joules, the heat energy released when 1 mole of sodium chloride is made. (2)

(Total for question = 8 marks)

Student 1 response Total 6/8	Marker comments and tips for success
a) 7 °C ✓	Correct answer calculated by mental arithmetic (no working shown).
b) 25 × 4.2 × 7 0 = 735 J ✓	1 mark only. A mark was lost for failing to give the equation for how to get the answer. Always write the formula before substituting the values. The calculation uses the wrong mass. You need to use the total volume of *all* the liquids and convert that into grams. The mass of the apparatus is usually ignored unless you are told this is the question. The mark is gained for correct calculation.
c) 1 mol × 25/1000 ✓ = 0.025 mol ✓	Correct. Remember that you find the number of moles by dividing the volume used by 1000 cm³ and then multiplying by the concentration of the solution.
d) 735/0.025 ✓ = 29 400 ✓	The answer is wrong, but the method is right. Both marks are gained under error carried forward as the wrong value from part b) is used. Although units are not required, it is a good idea always to give units in your answer.

Calculations 1

Student 2 response Total 1/8	Marker comments and tips for success
a) 21 – 28 = –7 °C O	Make sure you do subtractions like this the right way round. A simple slip that can cost a mark is easily corrected. The temperature went up, so the value should be positive.
b) mass × specific heat/temp. change O = 50 × 4.2/ –7 = –30 J	The formula is wrong. You need to learn the formula to gain marks. Although the calculation is correct, there is no error carried forward as the wrong formula is not a previous answer.
c) 1 × 25 × 1000 = = 25 000 moles O	The number of moles is correct, but you **divide** the volume used by 1000, not multiply. This is a common mistake. Make sure that you get the multiplication and division the right way round.
d) –30/25 000 ✓ = –1.2 – 03 O	There is an error carried forward mark for the first line. The final answer is just a copy of the display on the calculator and so is wrong. Make sure you can convert this calculator answer to -1.2×10^{-3} or -0.0012.

Practice questions

2 A student reacted 1.0 g of three different metals with 20 cm³ copper sulfate solution in an insulated test tube. The student recorded the temperature rise for each reaction.

The table shows the results obtained.

Metal	Temperature rise / °C
zinc	8
lead	0
magnesium	15

    a) Calculate the amount, in moles, of zinc the student used. *(2)*

    b) Calculate, in joules, the heat energy released during the reaction of the zinc and copper sulfate. Assume the specific heat capacity of the solution is 4.2 J/g/°C and the mass of 1 cm³ of solution is 1 g. *(3)*

    c) Calculate, in joules, the heat energy released when 1 mole of zinc reacts with an excess of copper sulfate. *(2)*

    d) How much heat energy was released by the test tube containing the lead? *(1)*

    e) Calculate, in joules, the heat energy released when 1 mole of magnesium reacts with excess copper sulfate. *(3)*

    f) Suggest one source of error in the experiment. *(1)*

4 Physical chemistry

3 A student wanted to find out the molar enthalpy change of combustion of ethanol.

The student used this apparatus.

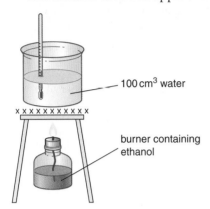

The student weighed the burner with the ethanol in before starting to heat the water. When the water's temperature had risen by 5 °C the student extinguished the flame and re-weighed the burner.

The table shows the results obtained.

mass of burner and ethanol at the start in g	75.20
mass of burner and ethanol at the end in g	74.28

a) Suggest **one** way the student could reduce heat losses from the apparatus. (1)

b) Calculate, in kilojoules, the heat energy released during the reaction. Assume the specific heat capacity of the solution is 4.2 kJ/kg/°C. (3)

c) Calculate the amount, in moles, of ethanol ($M_r = 46$) the student used. (2)

d) Calculate, in kilojoules, the heat energy released when 1 mole of ethanol burns. (2)

e) Burning ethanol produces water. Water can be split into hydrogen and oxygen. The equation for the reaction is: (2)

$$2H_2O \rightarrow 2H_2 + O_2$$

H-H bond energy/kJ/mol	O-H bond energy/kJ/mol	O=O bond energy/kJ/mol
436	464	498

Use information from the table to

i) Calculate, in kJ, the energy required to break the water molecules in the equation into H and O atoms. (2)

ii) Calculate, in kJ, the energy released when the H and O atoms form H_2 and O_2. (2)

iii) Calculate, in kJ, the enthalpy change for the reaction. (1)

4 When ammonium nitrate is dissolved in water a temperature change takes place.

A student dissolved 8 g of ammonium nitrate ($M_r = 80$) in 25 cm³ of water and recorded the temperature change.

Calculations 1

The table shows the results obtained.

Water temperature at the start / °C	Solution temperature at the end / °C
19	11

a) Explain whether the reaction is exothermic or endothermic. (2)

b) Calculate, in °C, the temperature change for the reaction. (1)

c) Calculate, in joules, the heat energy change during the reaction. Assume the specific heat capacity of the solution is 4.2 J/g/°C, and the mass of 1 cm^3 of solution is 1 g. (3)

d) Calculate the amount, in moles, of ammonium nitrate the student used. (2)

e) Calculate, in joules, the heat energy change when 1 mole of ammonium nitrate was used. (2)

f) A textbook showed that ΔH for the reaction was positive. What does this mean? (1)

5 A student investigated neutralisation. He read in a textbook that ΔH for neutralisation was always −57 000 J/mol for strong acids. The student decided to test if this was true.

a) What does ΔH represent? (1)

The student reacted 25 cm^3 of 1 mol/dm^3 hydrochloric acid with 25 cm^3 of 1 mol/dm^3 sodium hydroxide in a beaker. The molarity and volumes chosen enabled all the hydrochloric acid to be neutralised. The solution's temperature rose by 6 °C.

b) Calculate, in joules, the heat energy released during the reaction. Assume the specific heat capacity of the solution is 4.2 J/g/°C, and the mass of 1 cm^3 of solution is 1 g. (3)

c) Calculate, in joules, the heat energy released when 1 mole of hydrochloric acid is neutralised. (2)

d) Suggest **two** reasons why the figure you calculated in part c) is less than the value of ΔH. (2)

4 Physical chemistry

Calculations 2

Example

1 A student did a titration to neutralise some sodium hydroxide with some sulfuric acid.

The equation for the reaction is:

$2NaOH + H_2SO_4 \rightarrow Na_2SO_4 + 2H_2O$

A 25.0 cm³ sample of sodium hydroxide was neutralised by 23.50 cm³ of 0.100 mol/dm³ sulfuric acid.

a) Calculate the amount, in moles, of sulfuric acid used. (2)

b) Calculate the amount, in moles, of sodium hydroxide used. (1)

c) Calculate the concentration, in mol/dm³, of the sodium hydroxide solution. (2)

(Total for question = 5 marks)

Student 1 response	Total 3/5	Marker comments and tips for success
a) $0.1 \times \frac{23.5}{1000}$ ✓ = 0.0024 mol ✓		Correct. A clearly laid out answer showing how you should give an answer to 2 significant places. Don't waste time writing down every number on the calculator after the decimal place. Round up after the second or third figure.
b) 2 × 0.0024 = 0.0048 mol ✓		From the balanced equation and you can see that 2 moles of sodium hydroxide are needed to neutralise 1 mole of sulfuric acid.
c) $0.0048 \times \frac{25}{1000}$ = 0.0012 O		No marks. To calculate concentration of a solution from the moles used, the moles used should be multiplied by 1000 divided by the volume of solution used. It should read 0.0048 × 1000/25 mol/dm³

Student 2 response	Total 4/5	Marker comments and tips for success
a) $0.1 \times \frac{23.5}{100}$ ✓ = 0.00235 mol ✓		Correct. A clear answer showing 3 significant figures.
b) 0.00235 mol O		The answer is a guess without looking at the equation. Make sure you understand how equations show reacting amounts. The question asks for a calculation, so more is needed than simply copying a previous answer.
c) $0.00235 \times \frac{1000}{25}$ ✓ = 0.094 mol/dm³ ✓		1 mark for showing how to calculate the concentration of the solution, even though the 0.00235 moles used is the wrong answer from part b). This is an error-carried-forward mark. The second mark is for correct calculation.

Calculations 2

Practice questions

2 A student did a titration to find the concentration of some hydrochloric acid. The student measured the volume of hydrochloric acid needed to neutralise 25 cm³ of 0.200 mol/dm³ potassium hydroxide solution.

The equation for the reaction is:

KOH + HCl(aq) → KCl + H$_2$O

The diagrams show the readings on the burette at the start and finish of the titration.

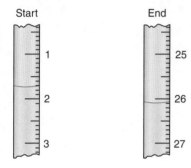

a) Make a copy of the table and use these diagrams to complete it, entering all values to the nearest 0.05 cm³. (3)

burette reading at end in cm³	
burette reading at start in cm³	
volume of acid added in cm³	

b) Calculate the amount, in moles, of potassium hydroxide solution used. (2)

c) Calculate the amount, in moles, of hydrochloric acid used. (1)

d) Calculate the concentration, in mol/dm³, of the hydrochloric acid solution. (2)

3 A forensic scientist wanted to find out the concentration of lithium carbonate in a water sample. The scientist titrated 25 cm³ of the lithium carbonate solution sample with 0.100 mol/dm³ hydrochloric acid. The scientist repeated the titration four times to check the results were repeatable.

The equation for the reaction is:

Li$_2$CO$_3$(aq) + 2 HCl(aq) → 2LiCl(aq) + H$_2$O(l) + CO$_2$(g)

a) Suggest **two** different methods of identifying when the reaction has been completed that the scientist could use. (2)

Here are the scientist's results.

	Trial 1	Trial 2	Trial 3	Trial 4
burette reading at the end (cm³)	41.30	39.25	40.10	40.60
burette reading at the start (cm³)	0.40	2.55	1.85	1.15
volume of hydrochloric acid added (cm³)	40.90	36.70	38.25	39.45

4 Physical chemistry

b) The scientist decided to use only three of the four results. Explain which result he discarded. (2)

c) Calculate the average volume, in cm³, of the three titrations the scientist used. (2)

d) Calculate the amount, in moles, of hydrochloric acid used. (2)

e) Calculate the amount, in moles, of lithium carbonate present. (1)

f) Calculate the concentration, in mol/dm³, of the lithium carbonate solution. (2)

4 A student made a salt by a precipitation reaction to find out the concentration of a solution of sodium iodide. He used the reaction of lead nitrate with sodium iodide.

The equation for the reaction is:

$2NaI + Pb(NO_3)_2 \rightarrow 2NaNO_3 + PbI_2$

The student placed 25 cm³ of the sodium iodide solution in a beaker. He added 0.200 mol/dm³ lead nitrate to the beaker until no more precipitate was formed.

The student filtered, washed, dried and weighed the lead iodide made. The mass was 9.22 g.

a) Explain why the lead iodide made was washed and dried. (2)

b) Calculate the amount, in moles, of lead iodide made ($M_r = 461$). (2)

c) Calculate the amount, in moles, present in 25 cm³ of the sodium iodide solution. (1)

d) Calculate the concentration, in mol/dm³, of the sodium iodide solution. (2)

5 A student carried out a titration to find the concentration of some sulfuric acid. She placed 40 cm³ of 1.5 mol/dm³ sodium hydroxide solution in a flask with an indicator. She placed the flask on a white tile and then titrated alkali with the acid. Her average titration value was 29.85 cm³.

The equation for the reaction is:

$2NaOH + H_2SO_4 \rightarrow Na_2SO_4 + 2H_2O$

a) Suggest a suitable indicator to use. Give a reason why you chose this indicator. (2)

b) Explain why the student placed the flask on a white tile. (1)

c) Calculate the amount, in moles, of sodium hydroxide solution used. (2)

d) Calculate the amount, in moles, of sulfuric acid used. (1)

e) Calculate the concentration, in mol/dm³, of the sulfuric acid. (2)

5 Chemistry in industry

Charges, chemical formulae and equations

Example

1. Aluminium is extracted from aluminium oxide (Al_2O_3), dissolved in molten cryolite at 950 °C. The diagram shows an electrolysis cell for the reduction of aluminium oxide to aluminium. Aluminium metal is formed at the negative electrode.

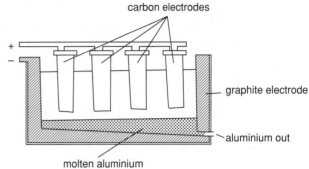

   The equation for the reaction is:

   $2Al_2O_3 \rightarrow 4Al + 3O_2$

   a) Name the substance formed at the positive electrode. Explain why it is formed. *(2)*

   b) Write an ionic equation to show:

   i) the reduction of the aluminium ion to an aluminium atom *(1)*

   ii) the oxidation of the oxide ion to oxygen. *(1)*

   c) One faraday is the charge of electricity that contains 1 mole of electrons. How many faradays of electricity are required to produce 1 mole of aluminium? *(1)*

   (Total for question = 5 marks)

Student 1 response Total 2/5	Marker comments and tips for success
a) Oxygen – it comes from the aluminium oxide. ✔	Oxygen gas is a product in the equation for the reaction, but you need to remember the details of aluminium extraction. The oxygen ion is attracted to the positive electrode, but as the electrode is made from carbon and is hot, the oxygen produced reacts to become carbon dioxide gas.
b) i) $Al^{2+} + 2e^- \rightarrow Al$ O	The charge on the aluminium ion is incorrect, so loses the mark. Always check the charges on ions. Using the Periodic Table, the group number of a metal is the same as the number of positive charges it has.
ii) $O^{2-} \rightarrow O_2 + 2e^-$ O	The charge is correct but the half-equation is not balanced, so the mark is lost. Use the Periodic Table to check the charge: 8 minus the group number gives the negative charge for non-metals. To balance the half-equation two oxide ions are needed on the left to produce oxygen gas, O_2, and four electrons are needed on the right.
c) 2 faradays ✔	Although the charge on the aluminium ion in the half-equation in b) i) is wrong, the 2 faradays gains the mark through error carried forward.

Student 2 response Total 5/5	Marker comments and tips for success
a) Carbon dioxide. ✔ The oxygen reacts with the carbon electrode. ✔	1 mark for the correct substance and 1 mark for a clear explanation. You need to remember that the carbon electrode is burnt away during the process.
b) i) $Al^{3+} + 3e^- \rightarrow Al$ ✔	The aluminium ion is correct, so the half-equation is easily balanced with three electrons.
ii) $O^{2-} \rightarrow \frac{1}{2}O_2 + 2e^-$ ✔	The equation is balanced by using a ½ in front of the O_2. This is an acceptable alternative to using two O^{2-} ions which avoids having to increase the number of electrons to 4.
c) 3 faradays ✔	It will take 1 faraday of electricity for each electron in the balanced half-equation for aluminium, so 3 are needed.

5 Chemistry in industry

Practice questions

2. Iron is a metal that is extracted by reduction with carbon, not by electrolysis.

   a) Explain why iron is extracted by reduction with carbon and not by electrolysis. (2)

   b) The equation for the reaction is:

   ___ Fe_2O_3 + ___ C → _____ + _____

   Copy and complete the equation for the reduction of iron oxide to iron. (2)

   c) Limestone, coke and air are added to iron oxide in the blast furnace. Describe the purpose of each of the three substances. (3)

   d) Suggest **one** advantage and **one** disadvantage of using carbon to extract iron from iron oxide. (2)

3. Zinc is a metal that can be extracted from its ore, zinc sulfide (ZnS). The zinc sulfide is reacted with oxygen to make zinc oxide, producing sulfur dioxide, and then reduced with carbon to zinc.

   a) Write a chemical equation of the reaction of zinc sulfide with oxygen to make zinc oxide. (2)

   The equation for the reduction of zinc oxide is:

   ___ + C → Zn + ___

   b) Copy and complete the equation for the reduction of zinc oxide to zinc. (2)

   c) Suggest **two** environmental problems that converting zinc sulfide to zinc may cause and explain how these problems can be overcome. (4)

   d) Explain why zinc oxide can be reduced by carbon to zinc. (1)

4. In the 1850s, Napoleon III of France often held banquets where the guests were given aluminium plates and cutlery to show his wealth. Aluminium was more expensive than gold.

   This was before a method was invented for making aluminium using electricity. All the aluminium made before the 1880s was made by reducing aluminium oxide (Al_2O_3) with expensive sodium or other very reactive metals.

   a) Write a chemical equation for the reaction of aluminium oxide with sodium to make aluminium. (2)

   b) Explain, using your equation, why aluminium was a very expensive metal in the 1850s. (2)

   c) Explain why carbon cannot be used to make aluminium from aluminium oxide. (1)

   Today aluminium is extracted by electrolysis. It is still more expensive than iron, but cheap enough to be used to make soft drink cans that can be thrown away.

   The diagram shows an electrolysis cell for the reduction of aluminium oxide to aluminium.

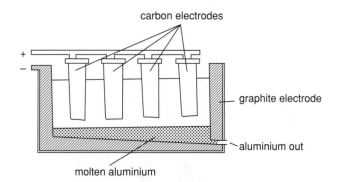

d) Explain why the aluminium oxide is dissolved in cryolite. (2)

e) Write the formula for the ion attracted to:

   i) the negative electrode (1)

   ii) the positive electrode. (1)

f) Write an ionic half-equation to show:

   i) the changes when an aluminium ion becomes an aluminium atom (1)

   ii) the changes when an oxygen atom becomes an oxygen ion. (1)

g) Explain why the half-equation shows that aluminium is an expensive metal to extract. (3)

5 Sodium metal is obtained by the electrolysis of molten sodium chloride.

The diagram shows a simple electrolysis cell for the production of sodium metal in a laboratory.

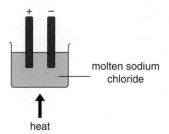

a) Write the formula for the ion attracted to:

   i) the negative electrode (1)

   ii) the positive electrode. (1)

b) Write an ionic half-equation to show:

   i) the reduction reaction (1)

   ii) the oxidation reaction. (1)

c) A scientist made 46 g of sodium metal. Calculate the number of faradays of electrical charge that was used. (2)

d) Suggest **two** safety precautions you should make, one for each product, if you attempted this reaction in a laboratory. (2)

5 Chemistry in industry

Practical work

Example

1. Some sodium chloride solid was dissolved in water. The solution was placed in a beaker with some universal indicator solution. Two inert electrodes were placed into the solution, and the solution was electrolysed.

   a) What colour would you expect the solution to be at the start? (1)

   b) i) Describe what you would see at the positive electrode. (1)

   ii) Name the gas produced at the negative electrode. (1)

   c) The solution around the negative electrode turned blue during the electrolysis.

   What does this show about the solution around the negative electrode? Suggest a substance that could be causing the blue colour. (2)

   d) The solution around the positive electrode was bleached during the electrolysis. Suggest a chemical that could be causing the bleaching. (1)

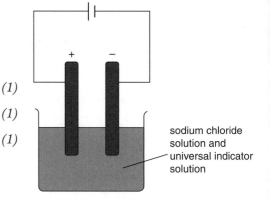

sodium chloride solution and universal indicator solution

(Total for question = 6 marks)

Student 1 response Total 3/6	Marker comments and tips for success
a) Colourless ○	The question has not been read carefully and the importance of the universal indicator has been missed. If an answer seems rather obvious, read the question again to make sure you've missed nothing.
b) i) Bubbles of gas ✔	The answer correctly describes what is seen at the positive electrode.
ii) Hydrogen ✔	The gas is correctly named.
c) The blue is caused by the universal indicator. ○	The answer shows a lack of understanding about what indicators are for. Universal indicator is used to monitor the pH, in this case the pH of the solution around the electrode to help identify the product. Read the question carefully and think about what each part of the procedure will do or how it might be useful.
d) Chlorine ✔	The gas is correctly named.

Student 2 response Total 4/6	Marker comments and tips for success
a) Green ✔	The mark is gained for realising that the universal indicator will provide a colour, and that as the solution is neutral it will be green.
b) i) Nothing ○	You should remember that you would see chlorine gas produced at the positive electrode in the electrolysis of brine.
ii) Hydrogen ✔	Correct. This answer is learned rather than understood. If the student had understood the process, they would have answered part b) i) correctly.
c) It is an alkali. ✔	1 mark for correctly identifying the reason for the blue colour of the universal indicator. The name of the substance is needed to gain the second mark. Make sure you suggest a substance if the question asks for one.
d) Chlorine ✔	The other gas is correctly identified.

Practical work

Practice questions

2 A student had done an electrolysis experiment using brine. She then watched a YouTube clip of the electrolysis of brine in industry. She took a screen grab of a diagram of the process.

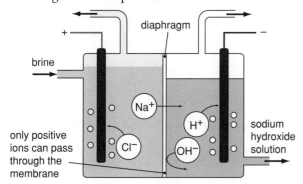

a) What is the chemical name for a solution of brine? *(1)*

b) i) Using information from the diagram, name the gas that is produced at the positive electrode. *(1)*

   ii) Describe a chemical test you could do to confirm the identity of the gas. *(2)*

c) i) Using information from the diagram, name the compound that is produced in the solution at the negative electrode. *(1)*

   ii) Write a half-equation for the formation of the hydrogen gas that is also produced. *(1)*

   iii) Describe a chemical test you could do to confirm the gas is hydrogen. *(1)*

d) What is the purpose of the diaphragm in the electrolysis cell?

3 A student did the electrolysis of some dilute sulfuric acid solution using the apparatus shown.

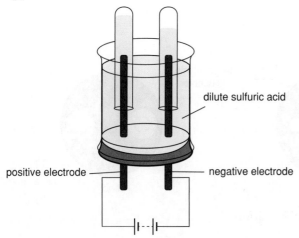

The volume of the gas given off at the negative electrode had twice the volume of the gas given off at the positive electrode.

a) Name the gas that was given off at the negative electrode. Describe a chemical test for this gas. *(2)*

b) i) The gas that was given off at the positive electrode was oxygen. Describe a chemical test for this gas. *(1)*

   ii) Write a half-equation for the production of this gas. *(1)*

5 Chemistry in industry

c) The student took the two gases and mixed them together. He burnt the gases, and collected the liquid that formed. Describe how the student could show the liquid was pure water. *(2)*

4 A student dissolved some sodium chloride in some water. He then electrolysed the solution in the apparatus shown.

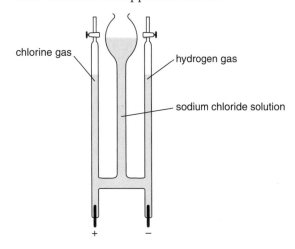

a) i) Describe a chemical test to show that the gas collected above the positive electrode is chlorine gas. *(2)*

  ii) Write a half-equation for the production of the chlorine gas. *(1)*

b) Describe a chemical test to show that the gas collected above the negative electrode is hydrogen gas. *(1)*

c) i) Sodium ions are attracted to the negative electrode where they gain an electron to become sodium atoms. Explain why sodium metal is not deposited on the negative electrode, and hydrogen gas is evolved. *(2)*

  ii) Write a chemical equation to show the reaction you have described in part **c) i)**. *(2)*

d) The products of the electrolysis of sodium chloride solution or brine are useful. Give **one** use for sodium hydroxide and chlorine. *(2)*

5 A student did the electrolysis of sodium chloride solution using a Petri dish. The student made up some sodium chloride solution, added some red litmus solution and placed two electrodes in the mixture. The two diagrams show the Petri dish from above, before and after the electrolysis.

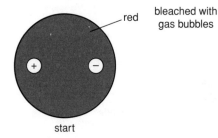
start

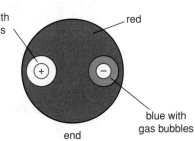
end

a) Copy and complete the results chart below. Some of the observations the student made have already been entered. *(3)*

	At the start of the experiment	During the experiment	At the end of the experiment
colour of solution in the centre of the Petri dish	red		
observations around the positive electrode		starting to go colourless	
observations around the negative electrode		changing from red to blue	
other observations	none	slight smell of chlorine	strong smell of chlorine

b) Explain why there was a strong smell of chlorine at the end of the experiment. *(1)*

c) Write a half-equation for the formation of the chlorine gas. *(1)*

d) The solution around the negative electrode is sodium hydroxide. Describe how it has been made. *(2)*

e) The gas at the negative electrode is hydrogen. Describe a test to show the gas is hydrogen. *(1)*

f) Suggest a safety precaution the student should take when doing the experiment other than wearing eye protection. *(1)*

6 A student electrolysed a solution of copper sulfate. He used platinum electrodes for the electrolysis.

a) Write a half-equation for the production of copper at the negative electrode. *(1)*

b) The positive electrode produced a gas. He found that the gas re-lit a glowing spill.

   i) Name the gas produced at the positive electrode. *(1)*

   ii) Write a half-equation for the production of this gas. *(1)*

c) Why did the student use platinum electrodes for his experiment? *(1)*

d) The student found that a mass of 0.635 g of copper has been deposited on the negative electrode.

Calculate the amount, in faradays, of electricity that was used to produce the copper. *(3)*

e) In a similar experiment using sodium chloride solution, 0.25 faradays of electricity were used. The half-equation for the production of chlorine gas is:

$$2Cl^- \rightarrow Cl_2 + 2e^-$$

   i) Calculate the amount, in moles, of chlorine gas that was made. *(1)*

   ii) Calculate the mass, in grams, of chlorine gas that was made. *(2)*

5 Chemistry in industry

Data analysis

Example

1 Crude oil is separated by fractional distillation into hydrocarbon fractions.
A hydrocarbon is a compound that contains only hydrogen and carbon.
A fraction is a mixture of hydrocarbons that contains molecules with a similar number of carbon atoms. The table gives some information about four fractions.

Common name of fraction	Average number of carbon atoms in the molecules	Average boiling point of fraction / °C	Time for ignition after flame applied / s
refinery gases	3	40	0
gasoline	8	110	1
kerosene	10	180	3
diesel	14	250	5

a) Explain why the table uses the average number to describe the number of carbon atoms in the molecules of a fraction. *(2)*

b) Describe the relationship between average number of carbon atoms in a fraction and average boiling point of the fraction. *(1)*

c) Explain why diesel is the most viscous of the fractions in the table. Use information from the table and your knowledge of hydrocarbons in your answer. *(2)*

d) Fuel oil and bitumen are two more fractions. Bitumen has more than 50 carbon atoms in its molecules and fuel oil has on average 35 carbon atoms. Explain which of the six fractions will be hardest to ignite. *(2)*

(Total for question = 7 marks)

Student 1 response Total 7/8	Marker comments and tips for success
a) Each fraction is a mixture of molecules. ✔	1 mark only. This is a statement of some knowledge, not an explanation. You need to link your knowledge with what is in the table, and why the average number is used, to get both marks.
b) The higher the boiling point the more carbon atoms. ✔	The answer correctly associates higher boiling point with greater number of carbon atoms, so gains the marks. However, the boiling point depends on the number of carbon atoms, not the other way round.
c) It has the largest number of carbon atoms.	The answer should refer to at least one piece of data from the table such as 'boiling point' or 'number of carbon atoms', and then explain how the chosen data relates to viscosity.
d) Bitumen ✔, it has most carbon atoms.	1 mark only. The explanation should link to the data in the table. Stating the trend of ignition would be a good way to start the answer before positioning bitumen as having most carbon atoms *in the molecule*, and suggesting from the trend that it would be hardest to ignite.

Data analysis

Student 2 response Total 7/7	Marker comments and tips for success
a) Each fraction is made up of lots of different molecules. ✔ The molecules have similar boiling points as they have similar numbers of carbon atoms ✔, so it's best to use the mean value for the carbon atoms.	A good answer using information from the table and question with a clear explanation of why is it useful to give the average number of carbon atoms for a fraction.
b) The higher the number of carbon atoms, the higher the boiling point. ✔	This answer clearly relates the boiling point to the number of carbon atoms.
c) Diesel has the highest boiling point which means the molecules are most strongly attracted to each other ✔. So the molecules will find it harder to slip past each other ✔ making it more viscous.	You should use the data in this way. Refer to the data, then use your knowledge to explain the information and its relevance to the question.
d) Bitumen. ✔ The trend from the table shows increasing difficulty to ignite with more carbon atoms, so as it is the largest ✔, bitumen is hard to light.	1 mark for the correct choice. The trend from the table is given and then used to support the answer.

Practice questions

2 A student burned samples of four different hydrocarbon fractions in the air. The student placed an inverted glass funnel at the same height above each burning fraction as shown in the diagram.

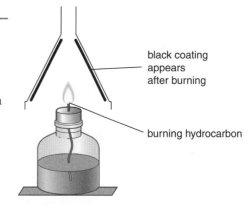

The student found the change in mass of the funnel after burning each hydrocarbon for 5 minutes. The funnel was cleaned after weighing before the next hydrocarbon was burnt.

The student's results are given in the table.

Crude oil fraction	Average number of carbon atoms in the molecules	Mass of black coating / g
gasoline	8	1.56
kerosene	10	3.9
diesel	14	8.52

a) i) Name the black coating that appeared on the funnel. *(1)*

   ii) Explain why this black coating appeared. *(1)*

b) Describe the relationship between average number of carbon atoms in a fraction and the increase in mass of the funnel. *(2)*

c) The student has failed to record one mass to the correct resolution. Which fraction was this? Explain your answer. *(2)*

d) Name the **two** products of the complete combustion of a hydrocarbon. *(2)*

e) Explain why the incomplete combustion of hydrocarbons in gas fires poses a health hazard. *(2)*

3 This question is about addition polymers. The commonest addition polymer is poly(ethene).

a) Describe the changes in bonding that occur when ethene forms poly(ethene). *(2)*

This table gives some information about three common addition polymers.

b) State **one** difference between the small molecules that make poly(ethene) and poly(chloroethene). *(1)*

c) Use a diagram to describe how the small molecules can join together to form poly(chloroethene). *(2)*

d) Teflon is a polymer that is used to give non-stick pans a slippery surface. The polymer is also known as PTFE.

77

5 Chemistry in industry

The repeat unit of a teflon polymer is shown below.

```
  F   F
  |   |
— C — C —
  |   |
  F   F
```

Draw the structure of the small molecule used to make this polymer. (1)

Polymer	Molecule the polymer is made from	Repeat unit of polymer
poly(ethene)	H\\ /H C=C H/ \\H	H H | | —C—C— | | H H

poly(propene)	H\\ /H
	  C=C—H
	H/ |
	        C—H
	        |
	        H
		H H
		| |
		—C—C—
		| |
		CH$_3$ H

poly(chloroethene)	Cl\\ /H C=C H/ \\H	Cl H | | —C—C— | | H H

e) If the T in PTFE stands for *tetra*, use the formula of the repeat unit to suggest the chemical name for PTFE. (1)

f) Explain why the disposal of addition polymers is difficult. (2)

4 Addition polymers are made by polymerising identical monomers. A different type of polymer can be made by reacting two different monomers together.

a) What is a monomer? (1)

b) Poly(ethene), poly(propene) and poly(chloroethene) are addition polymers. Identify a use for each polymer. (3)

The diagram shows a condensation polymerisation reaction between two monomers A and B.

[A] + [B] ⟶ —[A]—[B]— + H$_2$O

c) Explain why the reaction is called condensation polymerisation. (1)

d) Name a condensation polymer. (1)

5 The table shows some of the output of a fractional distillation column in an oil refinery and the local demand for each of the fractions.

Fraction	Percentage of the refinery output	Percentage demand locally
gasoline	15	26
kerosene	8	8
diesel	12	25
lubricating oil	26	14
fuel oil	29	20

a) Name the fractions for which demand is greater than the refinery's output. (1)

b) Explain why the figures in each column do not add up to 100. (1)

c) Cracking is a process that is used to make sure there is enough of each fraction to meet local demand. Describe what happens in cracking. (2)

d) Name the fractions in the table that should be used for cracking at this refinery. (1)

e) The alkane molecule $C_{20}H_{42}$ can be cracked to make a shorter-chain alkane and C_2H_4, ethene. Write a chemical equation for the cracking of $C_{20}H_{42}$ to make a shorter molecule and ethene. (2)

Working with graphs

Example

1 Ammonia is manufactured by the Haber process from nitrogen and hydrogen. The reaction is **exothermic**.

The graph shows the percentages of reacting nitrogen and hydrogen converted into ammonia at different temperatures and pressures.

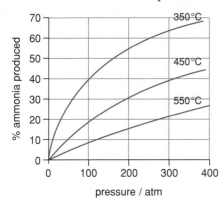

a) i) Use the graph to estimate the percentage yield of ammonia using a pressure of 300 atmospheres and a temperature of 350 °C. (1)

ii) Describe with reasons why increasing the pressure at 450 °C affects the yield of ammonia. (2)

b) Explain, how the graph shows that the reaction must be exothermic. (3)

c) In industry the Haber process uses 450 °C and 200 atmospheres pressure. Use your knowledge of the process and the graph to explain why these values are chosen when the yield is only 15–20%. (4)

(Total for question = 10 marks)

Student 1 response Total 2/10	Marker comments and tips for success
a) i) 20% ✗	The graph or question has not been read correctly. When you are asked to estimate or read a value off a graph, draw graph construction lines to help you read off the correct answer.
ii) Increasing the pressure increases the ammonia formed. ✓	1 mark only; the answer does not give a reason for the statement. To avoid missing parts of your answer, underline each part of the question that asks you to do something. You should underline *reasons* and *affects yield*, so you know to do both.
b) The graph gets higher at lower temperatures. ✗	Insufficient to gain a mark. The graph shows that less product is made as the temperature rises. In an exothermic reaction, heat is given out. This means that as the temperature gets hotter it is harder for the reaction to happen.
c) The high temperature lets the reaction go quickly, so too does the pressure. ✓	1 mark only; there is no reference to the graph. Use figures on the graph to explain your answer, for example 550 °C reaches 15% and 350 °C makes 60% much slower.

5 Chemistry in industry

Student 2 response Total 7/10	Marker comments and tips for success
a) i) 63% ✔ 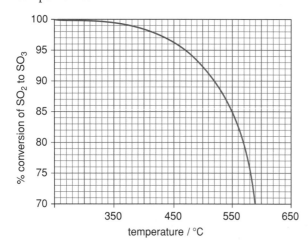	The answer is within the acceptable limits, and has been obtained by drawing lines on the graph to enable the value to be easily read off the graph.
ii) For each plot line at different temperatures, ✔ increasing the pressure increases the yield of ammonia. ✔	The answer clearly uses the data from the graph in the form of the three trend lines to explain how the pressure affects the yield.
b) As the temperature decreases the yield increases so the reaction is exothermic. ✔	1 mark only; there is no explanation of why this means that the reaction is exothermic. You need to comment on the backward reaction being endothermic and as such favoured by increasing the temperature, providing energy to be absorbed.
c) The graph shows the higher the temperature the less ammonia is made, but the reaction happens quickly. ✔ Increasing the pressure increases the ammonia made. ✔ The low yield is not a problem as the unreacted gases are recycled ✔ so eventually all become ammonia.	The answer uses the graph and refers to both temperature and pressure, with a concluding explanation. A reference to the need for speed to keep costs down would gain the final mark, as would a comment about competing priorities in maximising yield against time and costs.

Practice questions

2 Sulfur dioxide is made by burning sulfur in oxygen. The sulfur dioxide made reacts with more oxygen in the presence of a catalyst to make sulfur trioxide:

$$2SO_2(g) + O_2(g) \rightleftharpoons 2SO_3(g)$$

    a) i) Name the catalyst used to convert the sulfur dioxide to sulfur trioxide. *(1)*

        ii) Why is a catalyst used? *(1)*

        iii) What does the symbol $\rightleftharpoons$ in the equation mean? *(1)*

    b) The graph shows the percentage of sulfur trioxide produced at different temperatures.

Use the graph to answer these questions.

i) What is the percentage conversion to sulfur trioxide at 500 °C? (1)

ii) The reaction is carried out at 450 °C with a catalyst. What percentage of sulfur trioxide is made at this temperature? Explain why this temperature and a catalyst is used. (4)

iii) The reaction is exothermic. Suggest with reasons what would happen to the yield of sulfur trioxide at temperatures higher than 650 °C. (3)

3 Sulfuric acid and ammonia are used to make the fertiliser ammonium sulfate $(NH_4)_2SO_4$.

A student titrated 50 cm³ of ammonium hydroxide solution with sulfuric acid, to find out how much sulfuric acid was needed to neutralise the ammonium hydroxide solution.

The equation for the reactions is:

$H_2SO_4(aq) + 2NH_4OH(aq) \rightarrow (NH_4)_2SO_4(aq) + 2H_2O(l)$

The reaction is exothermic.

a) The student recorded the pH of the solution after each addition of sulfuric acid. The student stopped adding sulfuric acid after adding 50 cm³. Here is a graph of the results.

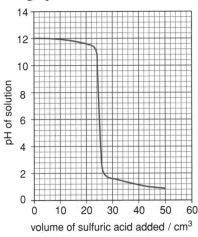

i) Describe in detail how the pH of the solution changed during the titration. (4)

ii) After what volume of sulfuric acid had been added was there most ammonium sulfate in the solution? Explain your answer. (2)

b) Use data from the graph to explain in detail how the student could obtain a pure sample of ammonium sulfate. (4)

c) What does the graph and equation tell you about the concentrations of the two solutions? (1)

4 The Haber process is used to make ammonia. The reaction is a dynamic equilibrium. This equation shows the reaction:

$N_2(g) + 3H_2(g) \rightleftharpoons 2NH_3(g)$

The forward reaction is exothermic.

a) What is meant by a dynamic equilibrium? (2)

b) Where are nitrogen and hydrogen obtained from for the Haber process? (2)

c) The graph shows how the percentage of ammonia formed varies according to the pressure of the gases and the temperature.

5 Chemistry in industry

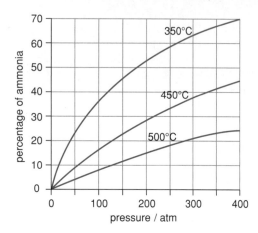

The reaction is usually done at 450 °C and 200 atmospheres pressure, with a catalyst.

Use the graph and your knowledge of reaction rates and equilibrium reactions to help you answer these questions.

　i)　What percentage of ammonia is produced with these conditions?　　(1)

　ii)　Explain why the reaction is carried with these conditions and a catalyst.　(4)

5 A class of students investigated how time affected the percentage of ethanol that could be produced by fermenting sugar solution with yeast. They set up three identical fermenting solutions and kept them at 30 °C, 40 °C and 55 °C. At the end of each day they measured the percentage of ethanol in each solution.

Here is a graph of the results.

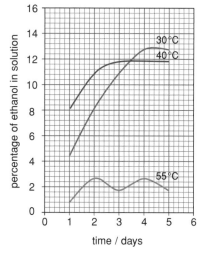

a) i)　At which temperature was most ethanol produced?　　(1)

　ii)　What was the maximum percentage of ethanol produced by any solution?　　(1)

　iii)　How long, in days, did it take the solution at 40 °C to produce its maximum percentage of ethanol?　　(1)

b) A scientist suggested that increasing the temperature increases the rate of the fermentation reaction. Using information from the graph explain whether you agree with the scientist.　　(4)

Calculations

Example

1 Limestone ($CaCO_3$) is used throughout the world to help farmers grow crops. Before the limestone can be used it must be changed to calcium oxide by heating.

The equation for the reaction is:

$CaCO_3(s) \rightarrow CaO(s) + CO_2(g)$

a) Calculate the mass, in grams, of 1 mol of calcium carbonate. (2)

b) The M_r of calcium oxide is 56. If 2 moles of calcium carbonate are heated and the carbon dioxide gas collected, calculate the maximum volume, in dm^3, of carbon dioxide gas that could be produced. (3)

c) A farmer heated 125 tonnes of calcium carbonate. All the calcium carbonate was converted to calcium oxide. Calculate the mass, in tonnes, of calcium oxide produced. (3)

d) Only 55 tonnes of calcium oxide was produced. Calculate the percentage yield of calcium oxide. (2)

(Total for question = 10 marks)

Student 1 response — Total 10/10	Marker comments and tips for success
a) $CaCO_3 = 40 + 12 + (16 \times 3)$ ✓ $= 100\,g$ ✓	Correct substitution of relative atomic masses and correct calculation of formula mass.
b) Reacting ratio is 1:1 so $1 \times 2 \times 24 = 48\,dm^3$ ✓✓✓	The equation shows that 1 mole of calcium carbonate produces 1 mole of carbon dioxide. If you heat 2 moles of calcium carbonate, 2 moles of carbon dioxide are made. You need to make it clear that the 1 is the reacting ratio in the equation and the 2 is from number of moles heated.
c) $\frac{125}{100} = 1.25\,moles$ ✓ so $1.25 \times 56 = 70\,g$ ✓ so 125 tonnes produces 70 tonnes ✓	Calculate the moles and mass that would be made first in grams, then scale up the mass from grams to tonnes. There is no mention of the reacting ratio in the answer. Here it is 1:1. Make sure you always show the reacting ratio from the equation even when it's 1:1 to avoid forgetting it.
d) $\frac{55}{70} \times 100$ ✓ $= 78.6\%$ ✓	Correctly uses answer to part c) to get the right percentage.

Student 2 response — Total 7/10	Marker comments and tips for success
a) $100\,g$ ✓✓	Correct. However, you should always show your working as a wrong answer may gain some marks for the working out.
b) $48\,dm^3$ ✓✓✓	It is unclear whether the calculation has taken into account the reacting ratio, but the answer is correct and gains full marks. It is risky to just write the answer as you either get full marks or no marks. Always show your working in case you make an error.
c) $\frac{125}{56} = 2.23\,moles$ ○ so $2.23 \times 56 = 124.9\,g$ ✓ so 124.9 tonnes ✓	The mass of calcium carbonate has been divided by the formula mass given in the question, which is for calcium oxide not carbonate, so the number of moles is wrong. The next calculation uses the wrong answer correctly, so gains 2 marks.
d) $\frac{56}{55} \times 100 = 102\%$ ○ $\frac{55}{56} \times 100 = 98.2\%$	The student has realised that 102% cannot be the yield so to get a figure less than 100 has switched the division round. The problem has been identified, but not the solution. If you get an unrealistic answer you should check the step before. Checking the answer to part c) would have corrected the problems in both parts of the question.

5 Chemistry in industry

Practice questions

2 Ammonia is an important industrial chemical. It is obtained from nitrogen and hydrogen gas.

   The equation for the reaction is:

   $N_2(g) + 3H_2(g) \rightleftharpoons 2NH_3(g)$

   a) What is the amount, in moles, of nitrogen gas needed to make 1 mole of ammonia gas? (1)

   b) Calculate the mass, in grams, of 1 mol of ammonia. (2)

   c) Calculate the volume at r.t.p., in dm^3, of 51 g ammonia gas. (3)

   d) Calculate the volume, in dm^3, of nitrogen gas that would produce 51 g of ammonia gas. (2)

3 Ammonia gas can be reacted with nitric acid to make the fertiliser ammonium nitrate.

   The equation for the reaction is:

   $NH_3(aq) + HNO_3(aq) \rightarrow NH_4NO_3(aq)$

   a) Calculate the mass, in grams, of 1 mole of nitric acid. (2)

   b) Calculate the mass, in grams, of 1 mole of ammonium nitrate. (2)

   c) A student reacted 0.2 moles of nitric acid with excess ammonia.

      i) Calculate the mass, in grams, of ammonium nitrate that should be produced. (3)

      ii) Only 5.6 g of ammonium nitrate was made. Calculate the percentage yield of ammonium nitrate. (2)

4 Iron metal is extracted from iron ore by heating the metal with carbon.

   The equation for the reaction is:

   $2Fe_2O_3(s) + 3C(s) \rightarrow 4Fe(s) + 3CO_2(g)$

   a) What is the name given to this type of reaction? (1)

   b) Explain why iron is extracted using carbon. (1)

   c) Calculate the mass, in grams, of 1 mol of iron(III) oxide. (2)

   d) Calculate the mass, in grams, of 1 mol of carbon dioxide. (2)

   e) If 8 g of iron(III) oxide is reacted with carbon:

      i) calculate the mass, in grams, of carbon required (3)

      ii) calculate the volume, in dm^3, of carbon dioxide produced. (2)

5 Electricity is passed through a solution of copper sulfate to obtain some copper metal. A current of 2.5 A was passed for 1 hour.

   The half-equation for the reaction is:

   $Cu^{2+}(aq) + 2e^- \rightarrow Cu(s)$

   a) Calculate the amount, in coulombs, of charge passed through the solution. (2)

   (charge in coulombs = current in amps × time in seconds)

   b) Calculate the amount, in moles, of copper deposited in 1 hour. (3)

   (96 500 coulombs of charge contains 1 mole of electrons)

   c) Calculate the mass, in grams, of copper deposited. (2)

   d) The electrolysis also produced oxygen gas at the positive electrode.

   The half-equation for the reaction is:

   $4OH^-(aq) - 4e^- \rightarrow 2H_2O(l) + O_2(g)$

      i) Calculate the amount, in moles, of oxygen gas produced. (3)

      ii) Calculate the volume, in dm^3, of oxygen gas produced. (2)

Longer-answer questions

Longer-answer questions

Example

1 Sulfuric acid (H_2SO_4) is made by a series of reactions between oxygen, sulfur and water.

   Describe how sulfuric acid is made. Your answer should include the reactants at each stage, the temperature and pressure used, and any catalysts that are used. *(6)*

   (Total for question = 6 marks)

Student 1 response — Total 1/6	Marker comments and tips for success
Sulfur is burnt with oxygen to make sulfur trioxide. O The sulfur trioxide is then <u>dissolved in dilute sulfuric acid</u> ✔ made by diluting concentrated sulfuric acid with water which then makes more concentrated sulfuric acid.	The answer is very brief. This part of the process is two stages, so no marks. There is no mention of temperature, pressure or catalysts, yet the question clearly asks for these. The dissolving of the sulfur trioxide in sulfuric acid is correct, but the sulfur trioxide should be dissolved in concentrated not dilute sulfuric acid. Underline the key prompts in the question, or write them at the start of your answer, then tick them off as you answer each one. The key prompts are *reactants, each stage, temperature, pressure, catalysts*. A brief plan before starting to answer is always a good idea. Write down as notes what you know of the stages, temperatures, pressures and catalysts before providing a detailed answer.

Student 2 response — Total 4/6	Marker comments and tips for success
Stage 1: Sulfur is burnt with oxygen to give sulfur dioxide. ✔	The answer is well structured. The key prompts of *reactants, each stage, temperature* and *catalysts* have been used. A balanced equation for stage 1 would help the answer, but the description in words is enough to gain the mark.
This is then reacted at <u>250 °C</u> and a <u>vanadium oxide catalyst</u> ✔ with more <u>oxygen</u> making sulfur trioxide ✔ (Stage 2): $SO_2 + O_2 \rightarrow SO_3$	The description of stage 2 gains marks for mentioning the catalyst and the need for more oxygen. Temperature is mentioned, but the value is wrong, so no mark. An equation is given, but it is not balanced.
In Stage 3 the sulfur trioxide is then <u>dissolved in concentrated sulfuric acid making more concentrated sulfuric acid.</u> ✔ Overall the reaction is: $S + O_2 + H_2O \rightarrow H_2SO_4$	1 mark for describing the final stage. The unbalanced equation shows an understanding of the overall reaction. The answer loses a mark because the key prompt of *pressure* has not been used. Check your answer against the key prompts and tick each one off as you find it in your answer.

5 Chemistry in industry

Practice questions

2 Crude oil is separated into useful fractions in a fractionating column.

  a) Describe how the separation of crude oil into useful fractions takes place in the fractionating column. (4)

  b) Some fractions are more useful than others. Describe how less useful fractions can be converted to more useful fractions, by cracking. (4)

3 Sodium hydroxide and chlorine are manufactured by the electrolysis of brine in a diaphragm cell. The diagram shows a diaphragm cell.

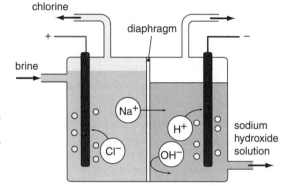

  a) Describe how the diaphragm cell produces sodium hydroxide and chlorine. Your answer should include the chemical names of the reactants, the purpose of the diaphragm, and any products not mentioned. (4)

  b) State a use of sodium hydroxide, and a use of chlorine. (2)

4 Ammonia is manufactured from the reaction between hydrogen and nitrogen.

The equation for the reaction is:

$N_2(g) + 3H_2(g) \rightleftharpoons 2NH_3(g)$

The percentage yield of the reaction is less important than the time taken to make some ammonia.

  a) Use your knowledge of the reaction to explain why the rate of the reaction is more important than the percentage yield of ammonia. (2)

  b) The reaction to make ammonia is exothermic.

  Describe the effect of raising the temperature and pressure on the equilibrium position, and why 450 °C and 200 atmospheres pressure is used. (4)

5 Aluminium is obtained by the electrolysis of purified aluminium oxide (Al_2O_3). The major cost of the process is the cost of electricity for the electrolysis and for heating. The aluminium oxide is dissolved in molten cryolite at 950 °C. The melting point of aluminium oxide is 2040 °C.

  a) Explain the economic benefit of dissolving the aluminium oxide in cryolite. (4)

  b) Aluminium and magnesium are metals that are produced by electrolysis.

  The half-equation for the reaction that produces aluminium is:

  $Al^{3+} + 3e^- \rightarrow Al$

  The half-equation for the reaction that produces magnesium is:

  $Mg^{2+} + 2e^- \rightarrow Mg$

  Use your knowledge of electrolysis and the half-equations to explain why it costs more to produce 1 mol of aluminium than 1 mol of magnesium by electrolysis. (2)

6 An accountant for a chemical company was very concerned to discover that the process his company used to make ammonia from nitrogen gas and hydrogen gas had a yield of only 15%. Using your knowledge of the Haber process, write an explanation for the accountant to help him understand that a 15% yield is highly cost effective. You should refer to the rate of reaction, temperature, pressure and the reactants and products. (5)

Index

acid rain 28
acids 48, 49, 52, 53, 54
activation energy 60
addition polymerisation 77–8
air 25, 26–7
alkalis 48, 49, 52, 53, 54
alkanes 40–2, 78
alkenes 40–1
aluminium
    extraction 69, 70–1, 86
    reaction with iron oxide 34–5
aluminium oxide 69, 70–1, 86
ammonia
    production 56, 57, 79–80, 81–2, 84, 86
    reaction with hydrogen chloride 8, 49
    reaction with nitric acid 84
    reaction with sulfuric acid 46, 81
    solution in water 8, 49
    structure 15
    test for 32
ammonium carbonate 32
ammonium chloride 8
ammonium hydroxide 8, 49, 81
ammonium ions 32
ammonium nitrate 64–5, 84
ammonium sulfate 46, 81
argon 6, 20
astatine 5
atomic number 4, 5, 17
atoms, structure of 4–6

baker's ammonia 32
barium 6
barium chloride 31, 52
barium sulfate 46
beryllium 35
blast furnaces 70
boiling point 1–2, 16, 17, 42, 76, 77
bonding 17, 41
Bordeaux mixture 32
brine see sodium chloride
bromide ions 32
bromine 5, 23, 34, 40, 41
butane 41
butene 41

calcium 6, 14, 20, 30, 35
calcium carbonate
    acid rain and 28
    formation 9
    heating 8, 83
    reaction with hydrochloric acid 29, 60
    reaction with nitric acid 50–1
    reaction with sulfuric acid 51
calcium chloride 15
calcium fluoride 14
calcium hydroxide 9
calcium nitrate 53
calcium oxide 8, 30, 47, 83
calcium sulfate 53

carbon
    electronic configuration 14
    forms of 18
    reaction with water 30
    reduction of metal oxides 70, 84
carbonates 29, 32
carbon dioxide
    electronic configuration 14
    production 29, 51, 69
    properties 29
    reaction with calcium hydroxide 9
    test for 28, 33, 51
    in water 8
carbonic acid 8
catalysts 43, 44, 57
    activation energy and 60
    effect on reaction rates 55–6, 57, 80–1, 82
changes of state 1–2
charge
    in electrolysis 69, 71, 75, 84
    on ions 13, 17, 19, 20, 27
chlorides 17, 24, 31
chlorine
    electronic configuration 12–13, 23–4
    manufacture 86
    reactivity 23, 34
    structure and properties 5–6, 17
    test for 32, 74, 75
collision theory 59, 60, 61
concentration 8, 66–8, 81
    reaction rates and 61
condensation 1–2
condensation polymerisation 78
conductivity 16, 18, 21, 24
copper 25, 32, 33, 45, 75, 84
copper carbonate 45
copper sulfate 11
    as catalyst 59–60
    electrolysis of 75, 84
    production 52, 54
    reactions with metals 63
    test for water 47
covalent compounds 15
cracking 78, 86
crude oil 44, 76–7, 78, 86

diamond 18
diaphragm cell 73, 86
'disappearing cross' experiment 61
displacement reactions 34–5
displayed formulae 39, 40, 41
distillation 2
dot and cross diagrams 12–15, 24, 39, 40, 41
dynamic equilibrium 56, 81–2

electrolysis
    of acidified water 11
    of aluminium oxide 69, 70–1, 86
    of copper sulfate 75, 84
    of sodium chloride 71, 72–3, 74–5, 86
    of sulfuric acid 73–4
electronic configuration 4–6, 12–15, 19–20, 22–4
empirical formulae 10–11, 38
endothermic reactions 57
enthalpy changes 64, 65
equilibrium 56, 81–2, 86
ethane 39, 40, 41
ethanol 43–4, 64, 82
ethene 40, 41, 43, 44
evaporation 1–2
exothermic reactions 46–7, 57, 62–5, 79

fermentation 44, 82
flame tests 31, 32
fluorine 5, 14, 23, 24
formulae, finding 10–11
fractional distillation 76, 78, 86
fractions 76–7, 78, 86

gases 1–3, 16
giant covalent structures 16
graphite 18
Group 1 metals 22, 23, 24, 35
    see also individual elements
Group 2 metals 35
Group 7 (halogens) 5–6, 23–4, 24, 34
    see also individual elements
groups (Periodic Table) 4, 5, 6
gypsum 31

Haber process 79–80, 81–2, 86
half equations 25, 69, 71, 73–5, 86
halide ions 52
halogens see Group 7 (halogens)
heat energy 47, 62–5
homologous series 39–42
hydrocarbons 11, 38–44, 76–8
hydrochloric acid
    reaction with calcium carbonate 29, 60
    reaction with lithium carbonate 67–8
    reaction with magnesium 9, 50–1, 59
    reaction with metals 33
    reaction with potassium hydroxide 67
    reaction with sodium hydroxide 50, 62–3, 65
    reaction with sodium thiosulfate 47, 61
hydrogen
    electronic configuration 14
    production 7, 9, 11, 59–60, 72, 73
    test for 32, 50, 73, 74, 75
hydrogen chloride
    reaction with ammonia 8, 49
    structure and properties 15, 24, 37
hydrogen peroxide 36, 55–6
hydrogen sulfide 14

indicators 48–50, 72
iodide ions 32

Index

iodine 5, 23, 34
ionic compounds 15, 17
ions
    in acids/alkalis 49, 50
    tests for 31–2
iron
    extraction 70, 84
    rusting 11, 26–7, 37
    test for ions 32
iron(III) oxide 11, 34–5, 70, 84
iron(II) sulfate 32
isomers 41

krypton 6

lead 11, 63
lead iodide 54
lead nitrate 68
limestone (calcium carbonate) 83
liquids 1–3, 16
lithium 23, 24, 35
lithium carbonate 67–8
lithium fluoride 15
litmus paper 49, 50

magnesium
    combustion 26
    electronic configuration 6
    extraction 86
    oxidation 27
    reaction with copper sulfate 63
    reaction with hydrochloric acid 9, 50–1, 59
    reaction with metal oxides 35
    reaction with sulfuric acid 7
    reaction with water 30, 35
magnesium chloride 9
magnesium oxide 10, 15, 27, 50–1
magnesium sulfate 7
manganese(IV) oxide 36, 55–6
manganese sulfate 11
mass, calculations of 7–9, 10–11, 45, 46, 53, 75, 83–4
mass number 4, 5
melting point 1–2, 16, 17, 18
metals
    properties 16, 30
    reactions with hydrochloric acid 33
    reactions with metal oxides 34–5
    reactions with metal sulfates 34
    reaction with water 22, 33, 35
    transition metals 19
methane 14, 39, 41
methyl orange 49, 50
molar enthalpy change 64
molecular formulae 11, 39, 40
monomers 78

neutralisation 49, 50, 65, 66–7, 81
nickel chloride 54
nitrates 54

nitric acid
    reaction with ammonia 84
    reaction with calcium carbonate 50–1
nitrogen 15
non-metals 20, 21, 30

oxidation 25, 27, 35, 69
oxides 11, 21
oxygen
    electronic configuration 14
    percentage in air 25, 26–7
    production 11, 36, 69, 84
    reactions with elements 30
    test for 32, 73

particles
    rate of reaction and 60
    in substances 1–3
percentage yield 47, 79–82, 83, 84, 86
Periodic Table 4–6, 19–21
periods (Periodic Table) 4, 6, 19
phenolphthalien 50
pH scale 21, 48–9, 51, 81
poly(chloroethene) 77–8
poly(ethene) 77–8
poly(propene) 78
potassium 23, 24, 32, 35
potassium bromide 32
potassium chloride 54
potassium hydroxide 67
potassium manganate(VII) 3
potassium sulfate 37
precipitation 52, 53, 54, 68
pressure
    effect on equilibrium 56, 86
    effect on yield 57, 79–81, 81–2
properties of substances 16–18

reaction rates 58–61, 82, 86
    catalysts and 55–6, 57, 80–1, 82
reactivity
    electronic configuration and 19, 20, 23, 24
    of halogens 34
    of metals 23, 24, 33, 34, 35
redox reactions 35
reduction 69, 70
relative atomic mass 6
relative formula mass 7–9, 32, 46
reversible reactions 47, 56–7
rubidium 23
rusting 11, 26–7, 37

salts, preparation of 52–4
saturated hydrocarbons 41
sodium
    electronic configuration 14, 23
    from electrolysis 71
    reaction with water 22, 35
    structure and properties 22, 23, 24

sodium bromide 23
sodium chloride
    electrolysis of 71, 72–3, 74–5, 86
    electronic configuration 12, 13
    properties 17
sodium fluoride 14
sodium hydroxide
    production 74, 75, 86
    reaction with hydrochloric acid 50, 62–3, 65
    reaction with sulfuric acid 66, 68
    uses 74, 86
sodium iodide 68
sodium oxide 21
sodium sulfate 53
sodium thiosulfate 47, 61
solubility 17
sulfates 31, 32, 52
sulfur 14, 26, 30
sulfur dioxide 26, 56–7, 80–1
sulfuric acid
    electrolysis of 73–4
    manufacture 85
    reaction with ammonia 46, 81
    reaction with calcium carbonate 51
    reaction with magnesium 7
    reaction with magnesium oxide 50–1
    reaction with sodium hydroxide 66, 68
    reactions with zinc 59
sulfur oxide 21
sulfur trioxide 56–7, 80–1
surface area 59, 60

Teflon 77–8
temperature
    effect on equilibrium 56, 86
    effect on yield 57, 79–81, 81–2
    reaction rates and 58–9, 82
titration 52, 66–8, 81
transition metals 19

universal indicator 50, 72
unsaturated hydrocarbons 41

volume, calculations of 7, 8, 9, 46, 83–4

water
    energy to break molecules 64
    reactions with metals 22, 33, 35
    structure 15
    test for 30, 47

yield 47, 57, 79–82, 83, 84, 86

zinc 59, 63, 70
zinc carbonate 27
zinc chloride 53
zinc oxide 27, 53, 70
zinc sulfide 70